THOMAS JOHNSON

Iter (1629) *and* Descriptio Itineris (1632)

THOMAS JOHNSON
Botanical Journeys in Kent & Hampstead

A facsimile reprint with
Introduction and Translation of his
Iter Plantarum 1629
Descriptio Itineris Plantarum 1632

Edited by
J. S. L. GILMOUR

THE HUNT BOTANICAL LIBRARY
PITTSBURGH, PENNSYLVANIA
1972

Published by
The Hunt Botanical Library
of the
Hunt Institute of Botanical Documentation
Carnegie-Mellon University
Pittsburgh, Pennsylvania

Library of Congress Catalogue Card Number: 79-173847

Number 3 of the Hunt Facsimile Series

Printed in Great Britain at the University Printing House, Cambridge
(Brooke Crutchley, University Printer)

Contents

Foreword

Technology, by which twentieth-century man has propelled his material advancement to a nearly unassimilable rate, has made a lasting impact on the world of arts and letters. It has done so through the technical improvement of photomechanical processes of printing, an advancement that has led to the production of a myriad of facsimile editions of rare or relatively unavailable earlier works.

The production of facsimile editions is limitless, controlled only by the economics of the situation. Regrettably, in the field of the plant sciences, many such editions have contributed no more than a new printing of an earlier work. In some, the title-page imprint identifies only the reissuing publisher, place of business, and date, deliberately ignoring the responsibility to advise the uninformed user that it is a reprint, and failing to identify the edition reprinted. In others, the work faithfully reproduces the original with no clue—or only a buried one—to the producer's identity; this is borderline, if not actual, forgery.

Publishers of facsimile editions use the opportunity simply to fill a lacuna; or, more conscientiously, to provide student and scholar alike with a combination in one work of the old (the facsimile) plus new studies based on it. To perpetrate the former is to be only a merchant. To accomplish the latter is to acknowledge a responsibility to provide more than was available before. The first represents little more than the greed of the entrepreneur. The second reflects the selfless considerations of the scholar.

From its inception as a center for biobibliographical and historical studies within the plant sciences, this Institute has included in its program the sponsorship and production of facsimile editions which provide the reader with more riches than were available from existing materials on the subject. The volume in hand is an example of this program.

The selection of Thomas Johnson's two works, his *Iter Plantarum* of 1629, and his *Descriptio Itineris Plantarum* of 1632, for this facsimile brings to modern view the earliest separately published accounts of botanical exploration in Britain—the locale of both being the county of Kent, with sections on Hampstead Heath near London. First published in Latin, they were included by Thomas Sherman Ralph in his publication of the collected botanical works by Johnson: *Opuscula omnia botanica Thomae Johnsoni*, printed in London by William Pamplin, in 1847 (see p. 8). This too was wholly in Latin, and is itself a rare work. Only now do these historically important and fascinatingly written narratives by one of England's most famous herbalists become available in both facsimile and English translation, augmented as they are by scholarly studies that include modern names for the plants and localities cited.

Thomas Johnson the botanist, physician, and apothecary, as Mr Gilmour reports, is better known for the publication in 1633 of his substantial revision of John Gerard's English-language *Herball* (1597). The deserved glory was earned but was never fully accorded him in his day, since his publisher shrewdly capitalized on Gerard's fame and name, and minimized, as have many subsequent historians, the very considerable contribution made by Johnson to that edition. One may only speculate as to what heights Johnson might have attained had he not been killed in battle before reaching his middle years.

The choice of John S. L. Gilmour as Editor of this volume of the Hunt Facsimile Series was a natural one, since it was he who brought to my attention the existence of Canon Raven's translation and invited consideration of it as a work for the Hunt Institute to publish. Although John Gilmour is known to many persons as an oracle and arbiter in matters of plant nomenclature, and especially as concerns cultivated plants, to others as the pragmatic but highly informed philosopher in matters of classification and taxonomy, and to still others as a student of the freethought movement in today's society, it is to be remembered also that at one time in his career he was one of Britain's most perspicacious students of local Floras—those volumes, large and small, which from the time of Johnson to the present have sought to document most meticulously man's knowledge of the plants of that great island, Britain. In his youth Mr Gilmour possessed one of the most complete private collections of such Floras then known. By virtue of his tenure as onetime Assistant Director of the Royal Botanic Gardens at Kew, later as Director of the Royal Horticultural Society's extensive garden at Wisley, and since 1951 as Director of the University Botanic Garden, Cambridge, John Gilmour has acquired a fund of knowledge of plants and the literature about them, and of the people and circumstances that have produced that literature. In the course of this he has enveloped himself with a truly worldwide coterie of staunch and devoted friends to an extent enjoyed by few of his contemporaries. Since 1961 he has served as the central figure of this Institute's international Advisory Committee, contributing beyond measure from his wealth of experience to the formulation of policies for our programs and activities.

It is small wonder that John Gilmour was conversant with these two now excessively rare works by Johnson. More than that, the late Canon Raven, translator of the Johnsons, had been a personal friend. Kent and Hampstead Heath, the terrains of Johnson's accounts, have been known to Mr Gilmour for better than half a century. It was only logical that he should be invited to plan and to edit with a free hand this present facsimile edition. The confidence placed in him for the execution of his editorship is here reflected most meritoriously. To John Gilmour, and to the team of distinguished contributors to this volume, I express my unstinted gratitude and my thanks for the completion of an assignment well done.

GEORGE H. M. LAWRENCE

Introduction

1 *Thomas Johnson, the* Iter *and the* Descriptio

J. S. L. GILMOUR

'No lesse eminent in the Garrison for his valour and conduct as a Souldier, than famous through the Kingdom for his excellency as a Herbarist, and Physician.' This thumb-nail sketch of Thomas Johnson, from a contemporary pamphlet, paints a vivid portrait of the author of the *Iter* and *Descriptio*, who died fighting for the King at Basing House in 1644. Unfortunately few details of Johnson's brief life have come down to us, but all that is known can be found in *Thomas Johnson, Botanist and Royalist*, by H. W. Kew and H. E. Powell (1932). This excellent biography makes it necessary for me to give only a very brief outline of Johnson's life, as a background to his work on the British flora in general, and the publication of the *Iter* and the *Descriptio* in particular. I will quote no references, as these are fully set out in Kew and Powell.

The place and date of Johnson's birth are not known for certain, but the available evidence points to Selby in Yorkshire, about 1600. The next twenty years are almost a complete blank, but in the Court Book of the Society of Apothecaries, under the date 28 November 1620, there is the following entry: 'Thomas Johnson bound to Mr Wm. Bell for eight yeares.' Exactly eight years later, under the date 28 November 1628, the Court Book records that 'Thomas Johnson apprentice to Mr Wm. Bell was examined and found sufficient was made free paid his fees and gave a spoone'. Johnson, therefore, became a 'free brother' of the Society of Apothecaries in 1628, and he followed the profession of an Apothecary for the remainder of his short life.

By the end of the following year, 1629, Johnson had already made a name for himself as a botanist. In May he visited Hampstead Heath, and later in the year he made the first of his Kentish journeys (July), followed by a further visit to Hampstead (August)—and he printed records of these journeys in his first publication, the *Iter*. Further, he contributed a laudatory Latin address to Parkinson's *Paradisi in Sole Paradisus Terrestris*, which indicates that he was already regarded as a man of some importance. Johnson's chief botanical friends at this time were John Parkinson, George Bowles of Chislehurst, and John Goodyer, who later was to become his closest collaborator in the study of the British flora.

During 1632, Johnson, now established in practice on Snow Hill, in London, must have been a very busy man. Not only did he undertake his second Kentish journey, recorded in the *Descriptio*, 1632, but he was also working on his new and greatly improved edition of Gerard's *Herball* (published in 1633), and on his translation of the works of the surgeon, Ambrose Parey (published in 1634).

In the summer of 1633 Johnson again visited Kent, and in late October or early November he published what must be regarded as his *magnum opus*, Gerard's *Herball*, 'Very much Enlarged and Amended by Thomas Johnson Citizen and Apothecarye of London'. This was reprinted in 1636.

In 1634 Johnson visited Bath early in the summer and, in July, accompanied by a distinguished band of other apothecaries, he made a journey to collect plants in southern England, including the Isle of Wight. The results of this journey were embodied in *Mercurius Botanicus*, 1634, which, as shown by Kew and Powell, can be regarded as Johnson's first steps towards a complete Flora of the British Isles. As an appendix to the *Mercurius*, Johnson printed *Thermae Bathonicae*, a description of the city of Bath and its warm springs.

The next important event in Johnson's study of the British flora was his journey to Wales in July and August of 1639, the results of which he incorporated in *Mercurii Botanici pars altera*, 1641, his last publication.

In August 1642 the King raised his standard at Nottingham, and Johnson joined the Royalist forces at Oxford early in 1643. In May he was created Honorary Doctor of Physic at Oxford University—one of the famous 'Caroline creations', though, as Kew and Powell comment, 'undoubtedly with more show of justification than in many of these cases'.

By October 1643 Johnson was almost certainly with Colonel Rawdon at Basing House in Hampshire, the seat of the Marquis of Winchester, which was held for the King until near the end of the Civil War; and it was on 14 September 1644 that 'Lieutenant Colonel Johnson Doctor of Physique, was here shot in the shoulder'; 'whereby contracting a Feaver he dyed a fortnight after'.

Such, very briefly, is a summary of Thomas Johnson's all too short life. What place can he claim in the history of the study of the British flora—and, in particular, what importance can be assigned to his first two publications, the *Iter* and the *Descriptio*, reprinted and translated in the present volume?

There is no doubt, I think, that Johnson must be regarded as the outstanding figure among students of the British flora between the herbalist-botanists of the sixteenth century (William Turner, John Gerard, and Matthew de L'Obel) and the great John Ray who published his first book (*Catalogus*

Plantarum circa Cantabrigiam nascentium) in 1660. Although Johnson was an apothecary, and paid considerable attention to the medicinal properties of plants, he undoubtedly had a more purely botanical interest in the British flora than any of his predecessors. His combination of botanical exploration with a study of the incredibly complicated literature of the previous hundred years built a firm foundation for the work of Ray, and of many others, that was to come.

Johnson's study of the British flora is embodied in his *Iter* (1629) and *Descriptio* (1632), in his edition of Gerard's *Herball* (1633 and 1636), and in the two parts of his *Mercurius* (1634 and 1641). Until Kew and Powell's biography appeared in 1932, William How's *Phytologia Britannica* (1650) was regarded as the first attempt at a British Flora, but Kew and Powell showed conclusively that the *Mercurius* was 'the first work in which the then known British plants were enumerated; separated from those of the Herball; and dealt with alone'—and that the *Phytologia* was largely copied from the *Mercurius*. The scale of Johnson's contributions to our knowledge of the British flora is shown by the fact that in the *Iter*, the *Descriptio*, and his edition of Gerard, approximately 120 species were recorded for the first time in Britain, while in the two parts of the *Mercurius* nearly 900 species were enumerated, of which about 50 were recorded for the first time.

Although the *Mercurius* was the first attempt at a British Flora, Johnson, in 1641, was already planning to produce, with his friend John Goodyer, a much more extensive descriptive and illustrated Flora of Britain, and it was a tragedy that his death at Basing House in 1644 prevented the carrying out of this project.

Johnson's place in the study of the British flora is a secure one. What contribution to his achievement can be assigned to the *Iter* and the *Descriptio*? In the first place, to anyone interested in Johnson, they have the charm of 'first publications', foreshadowing what was to come; but, against a wider background, they were the first separate accounts of botanical journeys in Britain and as such merit a place of honour in the bibliography of British botany. Other botanical journeys had of course been made in Britain, for example by de L'Obel and Gerard, but their results had been incorporated in comprehensive works, rather than in separate volumes devoted to the journeys themselves.

In the *Descriptio*, Johnson can also claim another 'first'. The section on Hampstead Heath, taken in conjunction with the *Iter* Hampstead lists, aimed at setting out all the species that had hitherto been recorded for the locality, and hence can be regarded as the first British Local Flora, the herald of a host of such publications, now covering nearly the whole of the British Isles.

Apart from the purely botanical value of their plant lists, the *Iter* and *Descriptio* give us very entertaining accounts of seventeenth-century journeyings in Britain; these have not hitherto been available in English, though summaries have appeared in Trimen and Dyer (1869), Hanbury and Marshall (1899), and in the *Phytologist* for 1848. In this last, which is a review of the 1847 reprint (see p. 8), there is an appreciative reference to the party's fondness for good food and drink, which stimulated the editor to insert, in square brackets, the pompous rebuke that though 'it is needful for them [Natural History Clubs] to feed', and 'it is perhaps excusable that little harmless follies are committed at or after feeding time...it is very bad taste, in all instances, to print and circulate these particulars'.

Johnson's *Iter* and *Descriptio*, then, must always stand high in the esteem and affection of those who care for the British flora, and it is a belated tribute to their author that, albeit nearly three hundred and fifty years after they were written, they can now be read in English for the first time.

2. *The Bibliography of the* Iter *and the* Descriptio

J. S. L. GILMOUR

Collations

The collations of the two volumes are as follows:

The Iter [For title-page see Facsimile, p. 29] 4to. A-B^4. Pp. [16]; no page numbers. For the signatures and the distribution of contents see Facsimile, pp. 29–43; $[B_4]^v$ is blank and is not reproduced. There are water-marks on the inside edges of A_2, A_3 B_2 and B_3. No printer's name is given, but the printer's device on the title-page is very similar to No. 379 in McKerrow (1913), except that the letters A.H., at the base of the device in McKerrow's reproduction are absent in the *Iter*. McKerrow lists this device as used by A. Hart of Edinburgh in 1613 and 1619, but with no subsequent history, and says that 'this is perhaps a cast ornament with the letters inserted'. It is unlikely in the extreme that Johnson employed an Edinburgh printer, and the removal of the letters A.H. also suggests that, in 1629, another printer was using this device, perhaps as well as Hart's heirs. We shall probably never know who printed the *Iter* for Johnson.

The Descriptio [For title-page see Facsimile, p. 71]. 12°. A–B^{12}, $[C]^2$. Pp. [8]+ [44]; pagination irregular. For the actual pagination, the signatures, and the distribution of contents see Facsimile, pp. 71–96; $[C_2]^v$ is blank and is not reproduced. There are water-marks on the fore-edges of [A_7], [A_8], [B_{11}], and [B_{12}]. In all the copies I have traced (see below), except the British Museum copy, the two leaves [C_1] and [C_2], carrying the illustrations, are bound after [B_{12}], as shown in the collation above. In the British Museum copy, however, they are bound at the beginning of the volume, immediately before [A_1]. In the Facsimile, which was made from the British Museum copy, they have been placed in their normal position at the end of the volume after [B_{12}]. The printer's device on the title-page is No. 283 in McKerrow (1913), who states that it passed from Isaac Jaggard to Thomas and Richard Cotes in 1627. The earliest use recorded by McKerrow is by R. Jones in 1592; later, in 1619, it was used by Jaggard on *The Merchant of Venice* Q2. I am indebted to Mrs R. S. Bromwich and to Mr Leonard Hector for helping me with the Welsh words on the device, 'Heb Ddieu [modern Welsh 'Dduw'] heb ddim', which are part of the family motto of the Pennants of Downing and Bychton, in the County of Flint. The conclusion of the motto is 'a Duw a digon', the whole motto meaning 'Without God without anything; with God – enough'.

Publication dates

The Kentish journey described in the *Iter* began on 13 July 1629, and the visit to Hampstead took place on 1 August of the same year. In the only two copies of the volume that I have traced (see below) no date or place of printing is given (see Facsimile, p. 29), but it has been universally assumed that it was issued in 1629. Ralph, in his 1847 reprint (see below), prints 'Londini 1629' at the foot of the title-page, and Kew and Powell (1932, pp. 1 and 22), in their transcription of the title-page, include the same phrase, presumably copied from Ralph's reprint; but unless Ralph used a third copy which I have not traced and which bore this imprint (he does not give the provenance of his original), the printing of this place and date on the title-page is entirely imaginary! There is no entry referring to the *Iter* in the Registers of the Company of Stationers of London for either 1629 or 1630 (see Arber, E., 1875–94, vol. 4), which is not surprising, as Johnson printed it 'for the sake of my friends' (Preface of *Descriptio*, see p. 101). However, as there was an interval of five months between 1 August and the end of December 1629, and Johnson states in the *Descriptio* that he described the *Iter* journey 'with a hurrying pen' (see p. 101), we must, I think, assume, in the absence of evidence to the contrary, that it was printed in 1629. Strictly speaking, however, the date 1629 should be quoted in square brackets.

The *Descriptio* bears the imprint 'Excudebat, Tho. Cotes. 1632', and there is no reason to doubt that this was the year of publication. There is no entry referring to the *Descriptio* in the Registers of the Company of Stationers of London for either 1632 or 1633 (see Arber, E., 1875–94, vol. 4), though other books printed by Cotes are registered.

Location of copies

After considerable search and enquiry, I have located only the following copies of the *Iter* and the *Descriptio*, all of which I have examined:

Iter. Magdalen College, Oxford (one copy); British Museum, Bloomsbury, London (one copy).

Descriptio. Magdalen College, Oxford (two copies); British Museum, Bloomsbury, London (one copy); Department of Botany, Oxford (one copy); Department of Botany, Cambridge (one copy).

If any reader knows of other copies, I would be very glad to hear of them.

The Iter. Magdalen College, Oxford. This copy is bound with other works from John Goodyer's library, which was bequeathed to the college by Goodyer at his death in 1664, and which was rightly described by Canon Vaughan (see Gunther, 1922, p. 197) as 'a most splendid legacy of botanical treasures'. Gunther (1922, p. 275) states that the *Iter* 'appears to have been a presentation copy to Goodyer'. For corrections to the text, probably by Johnson, and an MS. list of plants at the end of the volume, probably by Goodyer, see Appendix IV, p. 151.

The British Museum. This copy is separately bound in green half-calf marbled boards. There is no indication of ownership, other than the British Museum stamp, and there are no MS. corrections or additions.

The Descriptio. Magdalen College, Oxford (copy 1). Separately bound in vellum, from the Goodyer library, with Goodyer's crest. Gunther (1922, pp. 275–8) states that this 'appears to have been the author's own copy with his MS. index, afterwards extended by How and used in the preparation of his *Phytologia* (1650)'...It 'may have come into How's possession after the death of Johnson in 1644'. For notes on corrections, the MS. index, and How's MS. additions, see Appendix IV, p.151 (copy 2). This is also from Goodyer's library, bound with other works; there are no MS. corrections or additions.

British Museum. Separately bound in green half-calf marbled boards. There is no indication of ownership, other than the British Museum stamp, and there are no MS. corrections or additions. For the position of the illustrations see under Collations, above.

The Department of Botany, Oxford. Bound separately in rebacked calf; the leaves are very closely cropped, many of the last lines of text being cut away. There is no indication of ownership.

The Department of Botany, Cambridge. Separately bound in marbled boards, in a slip case. On the first two blank pages inside the front cover are inscriptions indicating (1) that the volume had belonged to Thomas Rawlinson (the bibliophile, 1681–1725, see *Dict. Nat. Biogr.*); (2) that in June 1719 Rawlinson lent it to Thomas Hearne (the antiquary, 1678–1735, see *Dict. Nat. Biogr.*); (3) that John Martyn (1699–1768; Professor of Botany at Cambridge, 1733–62) acquired it on 25 October 1727; and (4) that Martyn presented it to 'Bibl. hort. bot.' (i.e. the library of the University Botanic Garden, Cambridge, founded in 1761). There is the following note on the second blank page, written by Rawlinson: 'Legi quicquid Itineris fuit Rei herbariae nescius 14[?] Jan. 1719.'

This has been translated by R. E. Latham as follows: 'Knowing nothing of botany, I have read whatever concerned the journey. 14[?] Jan. 1719', i.e. Rawlinson had read Johnson's account of the journey in Kent, but did not feel competent to study the lists of plants. The note is headed 'C. & P.'. I am indebted to Mr John Sparrow, Warden of All Souls, Oxford, for informing me that this abbreviation appears in many of Rawlinson's books and indicates that he had 'collated' the volume and found it 'perfect'.

Reprints

In 1847 T. S. Ralph published a volume entitled *Opuscula omnia botanica Thomae Johnsoni...nuperrime edita T. S. Ralph*...Londini sumptibus Guliel. Pamplin. M.DCCC.XLVII. It was dedicated, in Latin, to Edward Forster and William Borrer, and contained reprints, without any introductory matter, of the following works by Thomas Johnson: *Iter*, 1629; *Descriptio*, 1632; *Mercurius botanicus*, 1634; *Thermae Bathonicae*, 1634; and *Mercurii botanici pars altera*, 1641.

In 1849 T. S. Ralph re-issued the sheets of the 1847 *Descriptio* reprint, with a new title-page, as follows: *Thomae Johnsoni Descriptio Itineris Plantarum... Nup. edit. T. S. Ralph.* Londini: Sumptibus Guliel. Pamplin. M.DCCC.XLIX. It is not at all clear why Ralph and Pamplin should have issued this second volume, containing the *Descriptio* only, so soon after the 1847 volume; possibly Pamplin had some surplus sheets of the *Descriptio*, and decided to use them in this way—but the 1849 volume cannot have had a very big sale!

3. *Thomas Johnson's Latinity*

R. E. LATHAM

Johnson, in the *Iter* and *Descriptio*, displays a fairly good command of Latin. Though he can quote Virgil or Hesiod for mock-heroic effect, there is nothing in his writing to suggest a wide knowledge of the classics outside his own medico-botanical field. The loose and straggling construction of his sentences, with an inclination towards unattached participles (e.g. on p. 33 of the *Iter*:[1] *ad Gravesend festinantes, ab illis inventae sunt plantae*), is closer to seventeenth-century English than to Ciceronian Latin. Since he writes in the humanistic rather than the medieval tradition, applying classical names to contemporary institutions, we have to rely on his own explanations, or on external evidence, to discover that *praefectus* means now 'constable', now 'mayor' (*Iter*, p. 35), that William Broad's *servus* (*Iter*, p. 31) was in fact his 'apprentice', and that the title *nauarchus*, applied to the mayor of Queenborough, denoted (probably) the rank of naval captain. Similarly, we are left to guess what sort of boat is intended by the terms *scapha* (*Iter*, pp. 31 and 40), *myoparo* (*Iter*, p. 37, *Descriptio*, p. 76) and *actuarium navigiolum* (*Descriptio*, p. 90).

In choosing to write in Latin rather than English, Johnson may have been to some extent parading his erudition and perhaps also appealing to an element of professional vanity and exclusiveness in the Latin-trained apothecaries for whom he wrote. But there was still a feeling in his time that Latin was, as we should put it, a more scientific language than English, and therefore appropriate to writings which, for all their good-tempered badinage, were intended to be a serious contribution to knowledge. The main advantage of Latin was, of course, that, at a time when English was very little known abroad, it gave the writer access to an international public; but it seems unlikely that works such as these would have had any substantial audience among overseas readers.

[1] The page numbers are those of the continuous pagination of this volume.

4. *Johnson's Plant and Animal Names*

J. S. L. GILMOUR

It has been our aim, in the translations of the *Iter* and *Descriptio*, to give the correct modern scientific names for all the plants and animals listed by Johnson, under pre-Linnean names, as having been found by him and his companions in Kent and on Hampstead Heath. This is not an easy task and, as will be seen, the identity of several of the species listed still remains obscure. The main problems that arise are as follows: (1) Johnson may have misidentified the plant he found and, although the name he gives clearly refers to a particular species, this may not be the species he actually saw. In these cases, a knowledge of what was likely to be growing in the locality in the seventeenth century may lead to a reasonably certain identification. (2) Johnson may have assigned a name from, say, Gerard to a particular plant, but it may be difficult or impossible to be certain what species Gerard had in mind, even if there is an accompanying illustration. (3) A group of closely related species may have been lumped under one name in the seventeenth century, but these may have subsequently been separated into two or more distinct species, and it may be difficult to be certain which Johnson actually saw.

The method of attacking these problems has clearly had to be a combination of a study of the relevant literature with an application of a knowledge of the flora of the areas he visited. It is very fortunate that the standard Floras of both the counties concerned (Middlesex—see Trimen and Dyer, 1869; and Kent—see Hanbury and Marshall, 1899) give a selection of pre-Linnean names under each species, a not very common feature of British County Floras. These two Floras, therefore, have provided most valuable first guidelines, though every effort has been made to check their identifications from other sources. It has also been very fortunate to have had the help of Dr Rose, Professor Richards, Mr Price, and Dr Vevers, all of whom know very well at first hand the plants or animals of one or both areas concerned, and of Dr Margadant, who has a great knowledge of old bryophyte literature. Nevertheless, errors of identification may still be present, and it would be valuable for these to be pointed out by readers able to do so.

The modern scientific names used for British plants are, broadly, those in Clapham, Tutin, and Warburg (1962), though, in a few cases, others have been preferred. Where there is likely to be doubt as to the species intended, synonyms are given.

5. *The Flora and Landscape of North Kent, 1629-1971*
FRANCIS ROSE

One must always remember that in the early seventeenth century the English landscape was in general far more open than today—the great enclosures of the late eighteenth and early nineteenth centuries had not taken place. Towns were all very small and compact, outside London itself, and there would have been much common grazing land, far fewer hedges, and many large common fields. In Kent, however, there were great differences from other counties. There always appear to have been many small farms, individually owned or tenanted, so hedges may have been more plentiful than in other counties. Even at this time, fruit growing, especially cherries, was important in North Kent in the area between Rochester and Faversham, and this must have produced a more closed and hedged-in landscape than elsewhere. The earliest large-scale map of Kent of the mid-eighteenth century indicates indeed a landscape not unlike that of today in much of Kent, with many enclosed fields and a distribution of woodlands not very different from the present time, though many of the existing larger woods of today appear then to have been far more extensive. Even in the early seventeenth century there is unlikely to have been much open acid heathland in North Kent on the relatively fertile soils of this region, though there was certainly far more heath than today on the Lower Greensand country south of the Chalk Downs. Johnson did not visit this Lower Greensand country on these journeys, and may never have visited it, as he regards Hampstead Heath as possessing plants not to be seen in the parts of Kent he knew. Probably there was far more chalk grassland, used as sheepwalk, at that time in parts of Kent; Johnson's account, however, gives little evidence that this was particularly extensive in North Kent, even in the chalk soil areas between Dartford and Rochester and in the Isle of Thanet. The 'Chalkdale' area at Dartford is singled out for attention as something clearly rather unusual in his eyes. The flora recorded between Gravesend and Rochester is one of arable land, hedgebanks and roadsides, and suggests that the countryside there was largely arable land then as now. In Thanet, again, few indications are given from Johnson's lists that the countryside was other than basically as it is today in the parts remaining unurbanized—the plants he records, away from the actual coastline, are almost without exception those typical of roadside verges, hedgebanks, and arable land. On the journey from Sandwich to Canterbury, one can still identify the parts of the route where Johnson made his various lists, so little has the essential character of the countryside changed. For example, the woodlands just east of Canterbury show up in the lists, and the flora today is little different from what it was then: even some of the rarities Johnson records, such as *Dianthus armeria*, persist on open sandy banks in these woods.

It is then a remarkable fact that none of the plants we can identify with some degree of certainty appears to have become extinct in Kent—and the vast majority still can be found in exactly the areas where Johnson records them. One could indeed comment that Johnson must have missed a lot, as many other interesting wild plants that he did not record occur today along his routes, and probably occurred in his time. If he had also explored the high chalk downs and the Weald to the south, he would no doubt have found far more species than he did.

The biggest changes affecting the flora, apart from the larger (but still not overwhelming) areas occupied by the towns, have been along the coast. The 'wilderness' between Grain and Cliffe which Johnson found so wearisome to traverse was, however, still almost uninhabited sheep pasture and reedswamps until the huge Grain oil refinery was developed in the 1950s, and apart from the refinery area itself the whole of that district still remains remarkably rural today. Sheerness town now occupies the area where Johnson studied coastal plants near Queenborough, but all his species, and far more, remain on the coastal beaches in the east of the Isle of Sheppey. The coastal flora near Sandwich remains more or less intact.

In short, the modern traveller who chooses to follow Johnson's footsteps in Kent will find the countryside noisier and more populated, and the air less clean along the high roads—but he will be able to see nearly every one of the species that Johnson saw.

Johnson found approximately 330 plants which had not been recorded in Kent before. Many of these, perhaps nearly half, were first records for Britain.

6. *The Flora of Hampstead Heath, 1597–1971*

P. W. RICHARDS

Hampstead Heath, now a public open space of Greater London, holds a special place in the history of botany, for there must be few areas of its size anywhere in the world of which the flora is better documented. Because of its acid, sandy soil it was never cultivated and, lying barely five miles (8 km.) northwest of St Paul's Cathedral, it was from the sixteenth century onwards a favourite collecting area for the apothecaries and botanists of London. The botanical history of the Heath in fact goes back some years before Johnson's time, as John Gerard had already recorded a considerable number of species from Hampstead in his famous *Herball or generall historie of Plants* (1597). In the two hundred years following the publication of Johnson's *Iter* in 1629 Parkinson, Merrett, Ray, Doody, Petiver, Buddle, Dillenius, Hudson and others added to the list of species known from the area. During the earlier part of the nineteenth century Alexander Irvine, H. C. Watson and others left published or MS. records of the flora. Trimen and Dyer's scholarly *Flora of Middlesex* (1869) and E. C. de Crespigny's *New London Flora* (1877) give an accurate and probably fairly complete picture of the flora as it was in the second half of the century.

In the twentieth century there are contemporary records of the flora of Hampstead Heath in Thomas J. Barratt's *Annals of Hampstead* (1912) which includes a section on the plants by James E. Whiting and another by Miss C. Garlick; the latter in particular provides a considerable amount of fairly precise information about the more interesting species. A year later appeared *Hampstead Heath: its Geology and Natural History* (1913), edited by T. J. Barratt *et al.*, prepared under the auspices of the Hampstead Scientific Society; it contains chapters on the vegetation by A. G. Tansley, the trees and shrubs by H. Boyd Watt and the flowering plants by W. A. Whitton. Finally there is D. H. Kent and J. E. Lousley's *Hand List of the Plants of the London Area* (1951–7) which includes a large number of records from Hampstead Heath, among them a number by the writer, who knew the Heath well as a schoolboy in the 1920s.

From this botanical information, and from other sources such as the paintings and sketches of artists such as John Constable, who lived at Hampstead for many years in the early nineteenth century, it would be possible to reconstruct in some detail the gradual change and impoverishment in the landscape and flora over a period of some 400 years.

Hampstead Heath covers the crest and upper slopes of a hill 443 ft. (135 m) high, which with the neighbouring Highgate Hill forms London's 'Northern Heights'. Both hills are capped with Lower Bagshot Sands, an Eocene deposit which also forms the substratum of the much more extensive heaths of western Surrey, eastern Berkshire and northern Hampshire. At the boundary between Bagshot Sands and the impervious London Clay which underlies the surrounding area, including a large part of London, there is a springline, and one of the springs is responsible for the small bog on the West Heath which is referred to by nearly every botanist who visited Hampstead and where many of the most interesting plants were found.

In Johnson's time, and no doubt for many previous centuries, Hampstead Heath was an open common probably used by the neighbouring villagers who gathered gorse and firewood and as copyholders had the right to pasture cattle on it.[1] Though there were woods in the neighbourhood of oak (*Quercus robur* and *Q. petraea**),[2] birch (*Betula* spp.), hornbeam (*Carpinus betulus*), whitebeam (*Sorbus aria*), wild service (*S. torminalis*) and other trees, perhaps including beech (*Fagus sylvatica**), part at least of the heath itself was open and was probably a mosaic of plant associations in which common species were *Agrostis canina**, *Calluna vulgaris*, *Erica cinerea*, *Nardus stricta*, *Pteridium aquilinum**, *Ulex europaeus* and *Vaccinium myrtillus*, with scattered bushes of *Juniperus communis*: similar vegetation exists on many commons in Surrey, Sussex and Hampshire today. On the West Heath bog, and apparently on a smaller scale elsewhere along the springline, there were 'wet heath associations' in which abundant or dominant species included *Erica tetralix*, *Eriophorum angustifolium*, *Molinia coerulea**, *Salix repens* and *Sphagnum* spp.* with such associates as *Drosera rotundifolia*, *Genista anglica*, *Menyanthes trifoliata* and *Viola palustris**.

This complex of plant associations had doubtless replaced the former oak woodland climax vegetation at some much earlier period, possibly in Saxon times or even earlier. Its maintenance as an open, more or less treeless area was dependent on human activities such as burning and the grazing of domestic animals.

The changes of the flora of Hampstead Heath over its 400 years of recorded botanical history depended on the vast increase in the human population in its immediate neighbourhood, and on the changes in land-use consequent on the transformation of Hampstead from a rural parish to a unit in the London conurbation. In 1871, after some forty years agitation, an Act of Parliament

[1] Copyholder is one who holds an estate in 'Copyhold', which is a kind of land tenure, of ancient origin, as recorded in the copy of the lord's manorial court-roll.

[2] Species marked * were not recorded by Johnson. It is curious that he makes no mention of *Pteridium* or *Molinia*.

was passed authorizing the Metropolitan Board of Works to buy the Heath from the lord of the manor and it became the first open space in England to be purchased for public use. This finally ended the enclosures for building plots and private gardens which for many years had gradually encroached on it. The Act also stopped the removal of sand, gravel, and turf which had become an increasing threat to its plant life. In 1888, when the parish of Hampstead was transferred from Middlesex to the newly created administrative county of London, it was still at the fringe of the built-up area and for some thirty years more it remained connected with the open country by a narrow corridor of woods and green fields. By about 1920 the connection had been broken and, with the adjoining woodlands (Ken Wood etc.), the Heath had become an island enclosed on all sides by suburbs which now extend for miles north, east and west.

In the course of the nineteenth century what was once a typical southern English common surrounded by woods and agricultural land thus became a city park, the playground of an enormous urban population. Hampstead Heath, together with Parliament Hill Fields and several other adjacent areas, now form a public open space of some 625 acres (2·53 sq. km.), administered by the Parks Department of the Greater London Council.

In spite of crowds of people, atmospheric pollution, and other adverse influences, a remarkable number of wild plants still survive, though the vegetation of large parts of the Heath now consists mainly of swards of sown grasses and planted trees, often of non-native species. Many of the plants recorded by Johnson are still to be found and some wild species such as *Pteridium aquilinum*, *Rubus fruticosus* agg., *Ulex europaeus* and, more locally, *Nardus stricta*, are still abundant. A number of Johnson's plants, such as *Genista anglica*, *Hydrocotyle vulgaris*, *Salix repens* and *Vaccinium myrtillus* also survive, but in very small quantity (D. H. Kent). Other species, e.g. *Campanula rotundifolia* and *Juncus squarrosus* (the latter not mentioned by Johnson but recorded by Ray in 1670) could still be found in 1927 and, if they have in fact now disappeared, probably lasted for some years longer. In 1927 *Calluna vulgaris* itself was also still present though extremely rare, but *Erica cinerea* does not seem to have been reported since 1912.

Most remarkable perhaps was the survival of the flora of the West Heath bog. The partial drainage of the bog in 1881 seems to have been responsible for the disappearance, some time between 1877 and 1911, of *Drosera rotundifolia* which Johnson had seen in 1629, but *Eriophorum angustifolium* was fairly plentiful in 1927 (P. W. Richards) and probably still survives (reported by D. H. Kent in 1966). *Erica tetralix* was apparently extinct in the bog by 1912, but survived in another part of the Heath. *Menyanthes trifoliata* was reported as 'nearly extinct' in 1913. Of the bryophytes which are so characteristic of such bogs, *Aulacomnium palustre*, *Calliergon stramineum*, *Gymnocolea inflata*, *Sphagnum cuspidatum* and *S. subsecundum* were still present in 1927.

It seems that changes for the worse had already begun in Johnson's time, for he remarks of *Osmunda regalis* which Gerard had found 'in the midst of a bog at the further end of Hampsteede Heath at the bottom of a hill adjoining to a small cottage' (doubtless the West Heath bog) 'of late it is all destroyed' (Johnson's *Gerard*, 1633, p. 1131). *Osmunda* appears to have survived in other parts of the Heath and in the adjoining Ken Wood it was recorded as late as 1813.

A study of the various published plant lists and of the last dates at which various species were recorded shows that comparatively few were lost before the last quarter of the nineteenth century, and there is no reason to think that the plant associations had changed in any important respects between Johnson's time and then. After that time extinctions, especially among the smaller non-woody species, were numerous. One factor which must have been unfavourable to heather and other low-growing, more or less shade-intolerant, species associated with it was the growth of trees. Whiting (in Barratt, 1912) refers to the planting of larch (*Larix europaea*) and 'fir' (*Pinus sylvestris*) on the West Heath, apparently late in the nineteenth century, but even more important may have been the natural spread of birch, a normal feature of the succession from heath to woodland which takes place when grazing is removed. The development of dense thickets of brambles, bracken, etc. would have a similar effect. It may be surmised that a great increase in birch took place at the time when the area ceased to be rural so that grazing by domestic animals (and possibly also by rabbits) ceased. Lack of grazing would also have encouraged brambles and other shrubby plants. In the last 30 years a similar invasion by birch and bushes has overtaken many other commons in the London district which 20 years earlier were open heaths dominated by *Calluna*, *Erica cinerea*, *Ulex europaeus*, *U. minor*, and grasses tolerant of poor acid soils. The disappearance of the lichens recorded by Johnson, and possibly some of his other plants, was doubtless due to atmospheric pollution. The loss of *Lycopodium clavatum* (last recorded 1865) was perhaps not due to urbanization, as it seems to have disappeared from many other localities in southern England at about the same time (F. Rose).

In the last 40 years or so the wear and tear on the vegetation of Hampstead Heath has been enormous. The ceaseless traffic has worn away much of the natural grass cover and eroded the surface soil, favouring the expansion of bramble thickets still further. The public authorities have attempted to deal with the problem by temporarily fencing off enclosures and sowing ryegrass and other non-native species; in addition, weeds have invaded the Heath in many places. That a substantial number of Johnson's plants still survive is a testimony to the remarkable powers of endurance of some wild species.

Ken (formerly Caen) Wood which adjoins Hampstead Heath may well have been visited by Johnson, though he did not mention it by name. Though publicly owned since 1922, it was previously in private ownership as part of Lord Mansfield's estate and, no doubt for this reason, its vegetation is now in some respects the least modified in the Hampstead enclave. Ken Wood still preserves a high canopy of tall old trees (*Quercus robur*, *Q. petraea*, *Fagus*, *Carpinus*, *Sorbus torminalis* etc.) and it perhaps retains something of the character of the ancient forest of Middlesex of which it is almost the last relic. Unfortunately the growth of planted *Rhododendron ponticum* and *Prunus laurocerasus* has suppressed most of the ground flora (formerly including *Vaccinium myrtillus*, *Convallaria majalis* and, possibly as a native, the very rare *Maianthemum bifolium*), but some interesting species still survive in Ken Wood including, in a boggy area, six species of *Sphagnum* and *Polytrichum commune* which are now extinct on Hampstead Heath itself (F. Rose).

I would like to thank Mr Douglas H. Kent and Dr Francis Rose for their kind help in writing this section of the introduction.

7. *Field Botanizing in 1629 and 1971*

FRANCIS ROSE

Perhaps the main difference from today lay in the difficulties of travel in the seventeenth century. English roads were not properly made up until the late eighteenth century, when macadam (a surface of small stone chips rolled or beaten in) was invented. Although travel on horseback or by foot was probably not too bad in dry summer weather on the chalky, sandy, or gravelly soils of North Kent, the generally poor roads, even near London, encouraged travel by boat down the Thames, rather than by road, to reach most of northern and eastern Kent.

There were of course no such things as pocket Floras; still less were there identification keys, and nomenclature was in a very confused and rudimentary state. Hence identification of unknown plants was very difficult in the field. Even in the library, the bulky existing Floras were difficult to use, though such good illustrations as existed, for example, in Gerard's *Herball*, must have helped considerably.

Johnson says that he collected specimens of any rarities that he found, but no herbarium of his is known to survive, though far earlier collections still exist on the Continent. Presumably, however, Johnson was experienced enough to know much of the then existing systematics by heart, and he may have identified largely in the field, as do many experienced botanists today; perhaps he only collected what he was unsure of, or took material for pharmaceutical use. Although pharmacology was the main ostensible aim of these early botanists, one is left with the definite impression that the scientific study of plant distribution was being carried out for its own sake; the recording of localities for medically useful herbs may in fact have been of lower priority, as localities are rarely specified exactly. One feels that Johnson and his friends, like most modern field botanists, loved plants for their own sake.

Though travel must have been far more tiresome physically, and perhaps even physically dangerous from highway robbers at times for those travelling alone, at least there were great compensations. The peace of the sky and the countryside was still unshattered and was to remain so for over 200 years. There were no car fumes, and virtually no air pollution. Plants very sensitive to air pollution, such as epiphytic lichens and mosses, must have abounded on every tree and fence. There were no sprays of pesticides or chemical fertilizers, and cornfield weeds must have been very abundant—probably embarrassingly so to the farmers. Indeed elsewhere (in Gerard's *Herball*, edited by Johnson in 1633) Johnson remarks that *Bupleurum rotundifolium* was so abundant in cornfields south of Dartford that it might 'well be termed the infirmities of them'.

The low population meant that there was far less pressure on the landscape—recreational use beyond the immediate environs of towns did not exist. Clearly this meant that strangers wandering about off the beaten highways were sometimes viewed with suspicion and even fear; witness the incident with the Mayor of Queenborough. However, it is evident that men of education, once they had established their credentials, were made welcome everywhere and provided with almost embarrassing hospitality; there was certainly no need to worry about booking in advance at inns as is often so necessary nowadays, especially in the summer. Inns seem to have been glad of custom. The reaction of ordinary working people to his party of botanists is not recorded by Johnson, but they must have been fairly friendly, to judge by the ease with which a lift was obtained in a brewer's cart.

8. *The Present Volume and its Contributors*

J. S. L. GILMOUR

Dr Lawrence, in his Foreword, has sketched the background of the choice by the Hunt Botanical Library of Johnson's *Iter* and *Descriptio* for inclusion in their series of Facsimiles. I would like to add a word about the history of the project, and the part played by the various contributors, of whom the leading ones are the late Canon Raven (the translator), and Dr Francis Rose, who is responsible for the great majority of the modern botanical names, and for two sections of the Introduction.

The story starts some twelve years ago, when I conceived the idea that a translation of Johnson's two little volumes was an essential contribution to the history of British botany. Canon Raven was the obvious choice for the translator and Dr Rose for the up-dating of Johnson's pre-Linnaean names and both kindly agreed to undertake their respective tasks. When they were completed, fortunately before Canon Raven's death in 1964, I discussed with Professor Betts the possibility of the volume being printed by the private press of the Department of Fine Arts at Reading University. This was agreed but, owing to various circumstances, the project hung fire for several years. It was then revived by the suggestion that the volume should include facsimiles of the very rare originals, as well as the translations, and that it should be published in the Hunt Facsimile Series. Reading University kindly agreed to give up any rights in the printing of the translations, and the Hunt Botanical Library asked me to edit the volume. The Cambridge University Press agreed to undertake the printing and, at long last, the enlarged project finally got off the ground!

In addition to Canon Raven and Dr Rose, as will be seen, we have been fortunate enough to enlist the help of a number of other specialists in various aspects of the volume. Mr Ronald Latham, the translator of Lucretius, has prepared Canon Raven's translations for final printing, has added a number of verse translations and footnotes, and has written the section in the Introduction on Johnson's Latinity. Professor P. W. Richards and Dr W. D. Margadant have tackled the difficult task of interpreting Johnson's names for bryophytes and lichens, and Professor Richards has contributed an historical account of the flora of Hampstead Heath; Mr J. H. Price and Dr H. G. Vevers have undertaken the equally difficult interpretation of Johnson's algal and animal names respectively, and Dr Maurice Burton has contributed a fascinating note on the stuffed 'sea serpent' seen by Johnson in Mr Duck's shop at

Sandwich. Dr Philip Gaskell has helped with bibliographical problems, and Mr Clive King, Mrs Nicola Hurst, Mr Nicholas Wall, and especially my daughter, Miss Clare Gilmour, have given me considerable assistance in various ways, including the preparation of the indexes. I would like to thank all these contributors for their helpful co-operation; and last, but by no means least, my grateful thanks are due to the Cambridge University Press, not only for their expected efficiency and expertise, but also for many valuable suggestions, and assistance with the maps—and to Dr George Lawrence for his constant patience, help, and encouragement.

Thomas Johnson's
Iter Plantarum, 1629
in facsimile

ITER PLANTARVM INVESTIGATIONIS

Ergo Suſceptum

A decem Socijs, in Agrum CANTIANVM. Anno Dom. 1629. *Iulii.* 13.

ERICETVM HAMSTEDIANVM

Sive

Plantarum ibi creſcentium obſervatio habita, Anno eodem 1. *Auguſti*

Deſcripta ſtudio, & operâ *Thomæ Iohnſoni.*

Iter Plantarum investigationis ergô Susceptum.

PAucis abhinc elapsis annis, consuetudo verè laudabilis inter rei herbariæ studiosos crevit, bis aut sæpiùs quotannis triduum aut quatriduum iter Plantarum investigationis ergô suscipere. Hoc anno societati visum est Agri Cantiani partem perambulare, diesque statutus est, (scilicet *Iulij* 13°.) cuius memores primo mane in D. *Pauli* Templo convenerunt Dominus *Ionas Styles*, *Gulielmus Broad*, *Iohannes Buggs*, *Leonardus Buckner*, *Jobus Weale Robertus Larking*, *Thomas Wallis*, duo *Edwardi Brownes*, (quorum alter servus *Gulielmi Broad*) & Ipsemet. Tum properantes ad fluminis ripam duas scaphas conducimus vsq; ad *Gravesend*, sed iam ripas linquentibus

Vir. Æn, 1. *Eripiunt subitò nubes cælumq; diemq;*
Nostrorum ex oculis: ponto nox incubat atra;
Intonuere Poli, & crebris micat ignibus æther,
Præsentem nobis intentant omnia mortem.

quâ de causâ scapha, in qua *Bucknerus*, *Buggs*, *Weale*, & *Larking*, Greenwichum divertit, illiq; cæli iniuria pertæsi terram appulsi se refecerunt. Nos ad *Gravesend* nullâ interpositâ morâ tendimus, & pransi ad *Rochester* (scripto in sociorũ adventum in hospitio relicto, quo in loco pernoctare decrevimus) solitâ viâ tendentes, invenimus plantas, quarum nomina sequuntur

Lichen è diversorij muro avulsus.
Cynoglossum maius vulgare.
Verbena.
Irio sive Erisimum Dioscor.
Hyosciamus niger Vulgaris.
Cicuta.
Cichoreum sil.
Perfoliata.
Ranunculus repens vulgaris.
Iacobæa maior & minor.
Elaphoboscum erraticum, sive pastinaca siluestris.
Euphrasea vulgaris,
Euphrasea rubra, sive secunda Dodonæi.
Rubia silvestris.
Hedera terrestris.
Ocimoides, Been album Monspelliensium, sive Papaver spumeum.
Medica semine racemoso.
Sedum, sive Sempervivum maius vulgare.
Sedum medium vulg.
Sedum tertium Dios. illecebra.

A 2 *Corono-*

Coronopus Ruellii Cornu cervi alterum vulgi. Lob.

Hordeum spontaneũ spurium. Holcus Plinii, Anguillaræ

Erigeron, senecio.

Bursa Pastoris,

Dens Leonis, Taraxacon, Rostrum porcinum Caput Monachi.

Solanum hortense

Solanum Lignosum, Dulcamamara, vitis silvestris,

Oxalis, acetosa vulgaris.

Oxalis ovina.

Sonchus aspera vulgi.

Sonchi Lævis species 2 aut 3s.

Lapathum acutum vulgare.

Lapathum acutum minimum (Lob.

Luteola

Horminum silvestre.

Daucus silnestris vulgaris.

Spondilium, Branca ursina Germanica.

Myrrhis silnestris

Lithospermũ sive milium solis.

Viorna. Lob. Ἀτραγένυ *Theophrasti. Clus. Hist.*

Argentina sive Potentilla.

Sagina spergula.

Thlaspi vulgare Mithridaticum, Vaccariæ folio, Lob

Caucalis silvestris.

Marrubium aquaticum acutũ sive Panax Coloni Gerardi.

Marrubium vulgare.

Ballote, siue Marrubium nigrum.

Crataegonon. Lob.

Carduus acaulis Lob.

Carlini silnestris Clusii.

Hieracium maius.

Hieracium chondrillæ facie.

Hieracium minus præmorsa radice.

Psendomelanthium. Nigellastrum, Lychnis segetum.

Cyanus minor vulgaris.

Antirrhinum medium

Ladanum segetum Lugd.

Bromos sterilis altera,

Aremisia vulgaris.

Papaver rhœas.

Tithimalus helioscopus.

Convolvulus minor purpureus & albus.

Helxine Cissampelos Atriplicis facie, Lob.

Esula exigua Tragi.

Scabiosa segetum.

Gnaphalium vulgare, herba impia Plinij

Elatine Dios. Veronica fœmina Fuchsii.

Elatine altera, Lob.

Iacea maior purpurea.

Cotula fætida.

Anchusa degener facie milii solis.

Rubia sil: purpureis parvis floribus, fortè stellaria. Lugd.

Perchpier Anglorum. Lob.

Saxifraga Anglicana, Alsines minimum genus Daleschampii polyganum selinoides Gerardi.

Pecten veneris, Scandix.

Rapistrum arvorum flo: luteo, &

& flore albo eruca foliis. Lob.

Iacea nigra vulgaris

Scabiosa tenuifolia.

Persicaria, Hydropiper.

Persicaria mitis maculosa,

Bryonia alba, vitis alba.

Bryonia nigra. Sigillum B. Mariæ.

Polygala flo: rubris, & flo: cæruleis, flos ambervalis Dod.

Lathyris Leguminosa maior Matt. Pisum Perenne vulgò

Alsine maior glabra.

Alsine media glabra,

Alsine hirsuta maior.

Alsine hirsuta minor.

Centaureum minus flo. albis & rubris.

Clinopodium vulgare.

Circæa Lutetiana. Lob.

Onobrychis, sive caput gallinaceum Belgarum.

Caryophillata.

Fragaria.

Viola purpurea.

Scorodonia, sive Salvia agrestis.

Betonica.

Crista Galli lutea.

Prunella.

Lycopsis Anglica Lob.

Helianthemum, a quibusdam Panax chironeum. Cisti species.

Nepeta vulgo. Mentha Cattaria.

Origanum onites Matt. Maiorana Anglica Gerardi, Hyssopus genuina Græcorum. Lob.

Pimpinella Saxifraga maior,

Pimpinella Saxifraga minor, sive Bipennella.

Demum pontem ſaxeum arcuato opere ſpectatiſſimum flumini Midweago, ibi rapido & torrenti ſuperpoſitum tranſeuntes ad Roffam (vulgo *Rocheſter*) pervenimus; & non multò poſt ad Hoſpitium noſtrum (quod Tauri inſigne habuit) ſequuti ſunt quatuor Socii Greenwichæ relicti; qui tonitru & imbre ſedato iterum ſe amni commiſerunt, ſed deficiente æſtu ad Erith ſcapham liquerunt, & terreſtri itinere ad Graueſend feſtinantes ab illis inuentæ ſunt, plantæ non à nobis obſervatæ.

Glycyrhiza silvestris Gesneri.

Glaux vulgaris, Lob.

Trachelium minus vulgare.

Typha maior vulgaris.

Thalictrum, sive Ruta pratensis Cordi.

Saponaria.

Tum Graueſend accepto noſtrorum deſceſſus indicio aſcenſis equis celeri curſu nos ad hoſpitium ſequuntur; ſic omnes læti vna cœnavimus, & in hoſpitis horto obſeruavimus plantas ſequentes.

A3 *Car-*

Cardiacam.
Hyoscyamum luteum.
Rosmarinum coronarium.
Ἀργυροκόμην sive gnaphalium Americanum Clusii
Hemerocallim Chalcedonicam Lob.
Chamaecyparissum
Hyssopum vulgarem sive Arabum.
Amaranthum purpureum.
Caryophyllum flo: albidis multiplicibus.
Delphinium flo: rubris & albis simplicibus
Pulmonariam maculosam.
Trifolium odoratum, Lotum odoram hortorum Lob.
Costum hortorum
Salviam.
Leucoium, sive Keiri flo: luteo simplici
Betam rubram.
Calendulam & tales vbiq; in nostratibus hortis cultas.

Sequenti mane ad *Chattam* iter dirigimus ad videndum classem omnium (vt *Camdeni* verbis vtar) quas sol videt bellico apparatu instructissimam,& in omnia momenta paratissimam, ex totâ classe vnam, quæ instar omnium esset, intravimus & lustravimus, *Principis Regalis* nomine navis sortita est, quæ tantum inter reliquas omnes eminet, quantum lenta solent inter viburna cupressi; Tormenta ænea maiora *66* numero habet, sturcturâ, magnitudine, magnificentiâ adeo clara, & præter omnium expectationem est, vt discribere nec audeo, nec, si ausus, possum. Classe relictâ, paulo infra Armamentarium Regium terram appulsi nobis se offerunt plantæ sequentes

Limonium copiose florens.
Glaux vulgaris.
Spargula marina Daleschamp. Anthylloides Thalij.
Absinthium marinum. Seriphium.
Plantago marina.
Serpentina Matthioli.
Caryophyllus marinus, sive gra. marinum.

Tum collem ascendentibus se offerunt

Linum sil: pusillum candidis floribus.
Acinos vera Anglica Clusii.
Anonis, siue Ononis,
Serpillum.
Trifolium corniculatum, sive Melilotus Coronata Lob.

Sic ad *Gillingham* venientes pransi sumus, & post prandium Cœmiterium ingressi vidimus

Ebulum copiose crescentem.
Parietariam, &
Filicem saxatilem Clusii templi muro avulsimus.

Nec

Nec præterire possum hospitalitatem (more patrio) ab Ecclesiæ Pastore D. Skelton nobis oblatam & acceptam; In cuius etiam domo invenimus viuidam effigiem, patrûm memoriâ, virtute, bonarumq; literarum studio clarissimi Equitis D. *Edwardi Hobæi*, cum hac inscriptione, *Sparsa & neglecta coegi*. In vnum enim magno & sumptu & labore nomina, insignia gentilitia, & viuas (quoad fieri potuit) imagines omnium Castelli *Quinborow Conestablium* (sic nos eius loci præfectum nominamus) coegit, & vltimo loco propriam posuit, quæ omnia temporum & sordidorum hominum injuria dispersa sunt. Hinc, gratijs redditis, pergimus vulgatâ viâ ad insulam *Sheppey*, & non ante observatæ plantæ conspectæ sunt

Agrifolium.
Taxus.
Smilax levis.
Sinapi agreste Apii, sive Laveris folio Lob.
Cornus fæmina
Bruscus.
Beta alba.
Cyperus rotundus inodorus septentrionalium Lob.
Gra. palustre Cyperoides Lob.
Polypodium.
Calamentha pulegii odore Lob.
Alsine foliis trissaginis.
Alsine foliis veronicæ.
Gra. tomentosum & acerosum Calamogrostis quorundam Lob.
Gra. tremulam minus vulg.
Lactuca agrestis odore opii.
Pilosella.
Orobanche, rapum genistæ,
Hypericon vbiq; in aridis.
Teucrium pratense.
Armeria pratensis, flos cuculi.
Armeria sil. altera caliculo foliolis fastigiatis cincto, &c. Lob. Cariophyllus montan°.
Lampsana, cum multis prius observatis.

Trajecto amne, & ingressa insula *Sheppey* nihil notatu dignum ad *Quinborow* venientes invenimus, ibi duobus hospitijs excipimur, postero mane dum iter aggressuri sumus, quidam venit & significavit, quod Loci Præfectus (nostrates *Maior* vocant) colloquium habere cum tribus aut quatuor nostrûm desideravit; annuimus: & ejus domum ingressi salutantem benignè resalutavimus. Tum ille nobis in hæc fere verba sermonem habuit. *Antiquis* hujus regni *Regibus* visum est, magna & lata Privilegia huic municipio concedere, vt tutius ab injuria nostram hanc parvam Insulam tueri possumus, ideoq; mihi (cui hoc tempore publica hujus loci salus

(creditur

creditur) datum est, vestri in hanc nostram Insulam itineris causam noscere; Non quòd aliquid in vobis video, quod suspicionis scrupuli aliquid moveat, sed quia talis semper nostra cura extitit, vt tantum hominum numerum, ignotâ causâ, hîc manere Majorum institutis non consentaneum est; Exponite ergo amicè vestri transitus consilium. Tum breviter *Iohannes Buggs*, (cui hoc negotij datum erat) Domine, causam nostri itineris quamvis vestrâ auditione vix dignam exponam; Rei & materiæ medicæ studio addicti sumus; eaq; de re hunc in locum pervenimus, vt rariores plantas hic in vestrâ insulâ crescentes investigemur, & non alia de causa tantum hoc spatium, à *Londino* huc vsque pedestri itinere confecimus. Tum Dominus *Styles*, præter relatam à socio causam, alia impulit, hominis tam bene merentis (vt tu es) videndi copiam habere, præcipuè cum in rebus marinis te adeo exercitatum esse noui (erat enim Regiæ Classis Nauarchus) maximè mihi jucunda semper tali cum viro familiaritas; Sicq; Præfectus his & talibus abundè satisfactus, breviter de rebus medicis & navalibus sermocinatus, cervisiâ optimâ oblatâ (solito more) nobis salutem precatus est, nosque eius humanitate freti gratiisque redditis vico discedimus, & in Castellum quondam munitissimum Regis *Edwardi* tertij opus nos conferimus; totum discribere non mihi in animo, nec præterire possum cœnaculum in quo non ita pridem totius agri Cantiani nobilium insignia gentilitia & nomina depicta sunt, in loco superiore per totius ambitum, *Vni soli & semper*, scriptum erat. In medio posita erant Insignia Regalia nunquam satis laudatæ principis *Elizabethæ*, & in eadem hæc carmina litteris majusculis scripta sunt:

Lilia Virgineum pectus, Regale Leones
Significant, vivas Virgo, regasque Leo:
Vmbra placet vultus, vultus quia mentis imago,
Mentis imago placet, Mens quia plena Deo.
Virgo Deûm vita, Regina imitata regendo,
Viua mihi vini fiat imago dei.

Qui Leo de Iuda est & flos de Iesse Leones
Protegat & flores, Elizabetha, *tuos.*

Dat. 1593.

Relict

Relicto Castro, in cuius summo fastigio Rutam murariam collegimus ad maris littus, ad solitumque opus properamus, & in itinere collecta sunt.

Halimus vulgaris Matt. Portulaca Marina.
Spartum nostras parvum.
Coronopus, stellaria, cornu cervi vulgi.
Kali maius, Salicornia,
Kali minus.
Crithmum Chrysanthemum.
Atriplicis marinæ species duæ

In ipso Littore crescebant.

Brassica marina monospermos Lob.
Papaver corniculatum flor. Luteo.
Cochlearia Anglica.
Tithimalus Paralius.
Alsines genus pelagicum Clusii, & vt mihi videtur Anthyllis peplios effigie. Lob.
Rubia marina.
Eringium marinum.
Kali minimum illecebræ facie
Tragum Matthioli.
Bryon lactucæ foliis Plin. Lob.
Lichen marinus.
Quercus marina.
Confervæ 2. aut 3. species.
Algarumq; 2. spe.

Et quæ non Plantæ, sed in materia medica comprehensa
Stella marina: & Ossa sepiæ, in littore collecta sunt.

Tum conducta Myoparone in Insulam Greane è regione nobis cursum dirigimus; Et Nauigiolo relicto 5. aut 6. milliaria ambulamus, nihil quod aliquid delectationis præberet, viso. Super ripas iter erat, in diei æstu miserâ site in mediis aquis, Tantali ritu, cruciamur (salsæ enim erant) nec minus premebat fames in inhumano illo tractu, vbi nec opidum in propinquo erat, neque fumus ad oculos, nec canum latratus ad aures, villarum solita indicia, perveniunt, quo languentes animos in aliquam spem erigere possumus; sed hisce tandem evictis difficultatibus, defessi ad Stoke venimus. Prandio parato & paracto, tota sociorum cohors iam languens (domino *Styles* & me exceptis) plaustrum cervisiarium ad Rochester itinerans conduxerunt. Nos ideo relictis sociis (quos in plaustro inter dolia morantes aurigarum curæ commisimus) a Stoke per Haslo superius, & Cowling oppidulum, & eiusdem nominis castrum iam ætate partim collapsum ad Cliffe tendimus. Et ne ignavia aut incuria nobis vitio verteretur, multas plantas partim non ante collectas in vnum conferimus, vt sunt.

B *Vicia*

Vicia sil: Tabernamontani, Aphacoides,
Chrysanthemum segetum.
Equisetum.
Anagallis flo: phœniceo.
Tussilago.
Lagopus.
Apium palustre in Stoke.
Sium terrestre.
Cannabis fæmina sativa.
Cannabis spuria, Lamium quorundam, Lob.
Eupatoreum Cannabinū mas.
Eupatoreum Cauuabinū fæm.
Genista vulgaris.
Genestella infectoria,
Lychnis sil. flo: albo.
Papaver spontaneum syl. Lob.
Carduus stellatus, Calcitrapa.
Carduus Polyacanthus Theophrasti.
Melilotum nostrum vulgare copiose.
Eupatoreum vulgare.
Holosteum Matt. Lob. Gramen ὀπηγονατόκαυλον *Thalii.*

Filix mas.
Filix fæmina
Cynoglossum maius vulgare
Tragopogon, barba hirci.
Vlmaria, barba capri, regina prati
Sium Matt, & Italorum.
Sium erucæfolium, Nasturtium aquaticum vulgi.
Arundo vallatoria.
Calamentha aquatica Belgarum, Lob.
Bellis maior, sive Consolida media vulnerariorum, Lob.
Personatia, Bardana, Lappa maior.
Lysimachia siliquosa maxima hirsuta.
Lysimachia siliquosa minor hirsuta.
Coniza maior helenitis. Lob.
Coniza media.
Sophia Chirurgorum.
Galeopsis vera Dios. Vrtica Herculea Taber.

Tum in *Cliffe* pernoctantes varios ab hospite rumores de socijs audivimus, postera mane, iter læti suscepimus, & in clivosis collibus nihil non prius observatum invenimus, quibus relictis, in salsis maritima non antea observata inventa sunt.

Tripolium maius & minus.
Althæa florens.

Trifolium fragiferum Clusii.

Et in agris Rapum sativum magnâ copia falce demessum, sic ad *Gravesend* solito hospitio revertentes sociorum plaustrantium adventum expectavimus, & cibum parari iussimus, prandientibus nuntius adfertur socios non longè abesse, parvâ interpositâ morâ adest *Tho: Wallis* conducto equo à *Rochester*

Rochester ad *Dartford*, & significavit paulò post se relictos esse *Leon: Bucknerum* & *Iohum Weale*, sed reliquos non ante vesperam venire ; sic dum nos eos expectamus, æstus relabitur, vt amnis beneficio eo die regredi non potuimus; inito consilio equos omnes cōparamus & Regia via ad *Dartford* properamus, ibiq; in Diuersorio quod Tauri insigne habuit equos discendimus, & statim ad locum multis & raris plantis refertum tendimus, *Chalkedale* vocatur, hic enim quondam lapides ad calcem vivam conficiendum eruti fuerunt, sed gramine pulcris variisq; plantis nunc ornatum invenimus, quarum nomina (omissis ante observatis) sequuntur

Gallium album.
Gallium Luteum.
Gentianella autumnalis, nondum florens.
Viola Calathiana Gerardi, trachelii species.
Rheseda, Eruca Cantabrica.
Pimpinella sanguisorba minor.
Baccharis Monspelliensum, Coniza maior Matt
Campanula minor rotundifolia
Centaurium minus luteum.
Scabiosa montana minima.
Anthillis leguminosa Belgarū
Orchis Myodes prima, Lob.
Serapias candido flo. montana Lob.
Buglossum sil. parvis floribus.
Buglossum echioides luteum.
Caucalis nodoso echinato semine Bauhini.

Vertendo ad dextram inuenta fuerunt

Erica vulgaris copiosis parvis floribus.
Iuniperus cum baccis.
Evonimus Theoph.
Blattaria flo: albis.

Sic ad oppidum reversi in ædes Domini *Wallis* divinæ gratiæ prædicatoris divertentes lautâ cœnâ benigne excipimur, & in eius viridario crescebant,

Ptarmaca vulgaris & *Bistorta maior*

Postero mane ex *Saltmarsh* habuimus,

Anagallidem aquaticam tertiam, Lob.
Allium sil. tenuifolium.

Hic pransi ad *Erith* tendimus, ædibus antiquis Dominæ *Dakins* ad sinistram relictis, in paludosis inuenta sunt,

Lysimachia purpurea spicata.
Lysimachia galericulata.
Valeriana sil maior.

B 2 *Iuncus*

Iuncus floridus, *Gladiolus aquaticus*, *Dod*

Sparganium ramoſum.

Trifolium paludoſum.

Alectorolophos, *Plin* *ſive pedicularis rubra.*

Acorus noſtras paluſtris cum ſemine.

Anagallis aquat. Latioribus foliis, flo. cæruleis.

Anagallis aquat anguſtioribus foliis, flo. albidis.

Myoſotis Scorpoides.

Hydrolapathum magnum.

Hydrolapathum minus.

Marrubium aquaticum, ſideritis prima Matt.

Betonica aquatica.

Plantago aquatica.

Hiſce locis relictis in vulgari tramite obſervata ſunt.

Helenium, ſive Enula Campana.

Pes Columbinus, geranium 2. Dioſ foliis amplis valdè diſſectis.

Blattaria vulgaris flo. Luteo,

Sic ad Erith venientes, ſcaphâ conductâ, reuertentes na-
ves tres ex India Orientali reverſas vidimus, earumq; vnam
ad Blackwal curſum dirigentem aſcendimus, *Bucknerusq;* nu-
ce indicâ maiore, cannâ fractâ & calamo Indico donatus fu-
it. Relictâ nave per pontem Britannico orbe celeberimum
ferimur, & ſtatim nobis relatum eſt, Socios eſſe reverſos &
quo in loco erant. Feſtinamus, inuenimus, & ab ijs intelligi-
mus poſt deceſſum noſtrum apud Stoake, Illi in plauſtro
magnificè per Ho ad Rocheſter venerunt, & in vrbis Præto-
ris domo (vbi & prius diverſabamur) pernoctabant; erat
Prætori nomen *Allen* qui nunc multò magis quam ante eos
humaniter excepit, & ampliſſimo cœnaculo honoravit; ſe-
quente mane ad ſalutandum adſunt dominus *Larkinge* Eccle-
ſiæ Roffenſis Præbendarius & Dominus *Selbeius* ejuſdem loci
generoſus, quibus pro humanitate ſuâ multum debemus. Re-
licta Roffa, in altera Fluminis Medweagi ripâ, in Stroad ite-
rum benignè invitati à D. *Mapeſton*, & D. *Mooreland* eius
loci generoſis, & *Iohanne* at *Noke* de *Cliffe* ad Agnum diuerſi-
modo paratum edendum, anuunt; ſic gratiis ijs pro beneficio
actis (vt etiam domino *Pitchard* rei herbariæ ſtudioſo, qui
illis Lunariã racemoſam mittere promiſit) directè ad *Grauęſ-
end* tendunt, & poſtridie ſolitâ viâ *Londinum* revertunt, vbi
iterum coniuncta ſocietas ſtatuit, *Auguſti* primo peragrare
Ericetum Hampſtedianum (vulgo *Hampſted Heath*) quoni-
am multæ ibi creſcunt, quæ hic deſiderantur.

Ericetum

Ericetum Hampstedianum.

STatuto tempore mane convenerunt septem ex decem socijs desideratis *Buggs, Weale,* & *Wallis*, sed in eorum locum succedunt *Iohannes Sotherton, Iohannes Marriott, Thomas Cresse* & nostri societati se jungunt, cœlo quamvis pluvioso non deterremur ad omnia parati, sed, qui maiora tulimus, minori cedere turpe duximus. Sic relictâ vrbe ad *Kentish-towne* cursum dirigimus, inde festinantes, sed non longe, progressos imber largus subsidium in *Highgate* petere coegit: sed imbre vix sedato moræ impatientes ad sylvam venimus, & a nobis plantæ non in priore itinere observatæ (illas enim consultò omitto, ne omnium enumeratio tædium afferret) sunt sequentes

Gram. spica Briza maius Bauh.
Tithimalus charachias amygdaloides.
Serratula flo pur. & albis.
Anagallis flo: luteo.
Astragalus Sylvaticus Thalij
Alnus nigra, frangula.
Sorbus sil: Alpina Lob. Ornus Dod.
Morsus Diaboli.
Pulmonaria Gallorum Hieracij flore vt mihi videtur.
Tormentilla.
Cynocrambe.
Ascyron.
Populus Lybica.
Betulus, Lob. Carpinus, Matt.
Virga aurea Villanovani.
Gnaphalium foliis longioribus.
Numularia.
Scrophularia maior.
Angelica silvestris.
Trifolium acetosum, Oxys Plin. Alleluia sive luiula Offic.

Relictâ sylvâ in ipso Ericeto provenientes conspecta sunt

Erica pumila calyculato vnedonis flore.
Erica Iuniperifolia altera Lob
Vaccinia nigra.
Lilium convallium.
Genistella aculeata.
Genista aculeata.
Lonchitis.
Muscus clauatus. Lob. Muscus terrestris, Matt.
Lysimachia galericulata minor, Gratiola latifolia, Ger.
Ornithopodium perpusillum.
Gram: Sparteum capillaceo folio minimum.
Clymenon Italorum, Androsæmum quorundam.
Ros solis.
Veronica mas vulgaris. Betonica Pauli.
Veronica pratensis, Dod.
Veronica recta minima. Lob.
Adiantum aureum majus.
Adianthum aureum minus.

B 3 *Muscus*

Muscus pyxadatos alablastriculos imitatus Lob.
Salix humilis bombicifera.
Iuncus bombicinus.
Cotiledon aquatica acris.
Ranunculus rotundifolius, fortè Apium risus
Ranunculus aquaticus hepatica facie Lob.
Ranunculus flammeus minor
Digitalis purpurea.
Aria Theophrasti effigie Alni.
Sorbus torminalis.
Alsines minimum genus peculiare non discriptum.

Ericeto iam relicto, in oppidulo *Hampstediano* nos paululum refecimus, et tum ad Kentish-towne revertentes, pransi sumus, et in pratis, semitis, et sepibus, ab euntibus et revertentibus observata sunt

Atriplex sil: sinuata.
Atriplex sil: Laciniatis folijs. Pes Anserinus.
Atriplex sil. Polygoni aut Helxcines foliis, Lob.
Aracus, sive Cracca maior & minor, Lob.
Bonus Henricus sive tota bona
Potomogeiton.
Arum baccis onustum.
Geranium Robertianum. Herba Roberti.
Geranium secundum Dios. Pes columbinus.
Geranium arvense Tab. Myrrhida Plinij.
Saxifraga Anglicana facie seseli pratensis.
Holosteum Ruellij, gramen Leucanthemum minus.
Barbarea.

Non omnino à proposito esse iudico, nomina quarundam plantarum hisce locis (primo die Maij) à nobis inventarum denotare, sed nunc partim aut prorsus arefactæ, quales sunt,

Anemone sive ranunculus nemorum flo: albo.
Hyacinthus vulgaris Anglicus.
Primula pratensis.
Primula silvarum.
Cardamine.
Lamium Luteum.
Bugula.
Alliaria.
Caltha palustris.
Chelidonium minus.

Sicque breviter, quicquid (omissis paucis vulgaribus) in hisce nostris perambulationibus conspectum fuit, in unum collegi, ut amicis, et alijs harum rerum cupidis innotescat, quantum

quantum in laboribus, quantum in ſumptibus, ad rei herbariæ ſtudium promovendum fuimus, et quod omnibus manifeſtum ſit, non ad oſtentationem, ſed ad uſum hi noſtri conatus fuerunt, diſcerpſimus, et repoſuimus quicquid rari occurrebat. Sed huius anni labores ſunt tanquam præludia alijs ſequentibus annis exantlandis, quibus propitius ſit Deus.

Amen,

FINIS.

Translation of Thomas Johnson's Iter Plantarum, 1629

By
CHARLES E. RAVEN
assisted by
RONALD E. LATHAM

Modern Scientific Names by
FRANCIS ROSE
assisted by
J. S. L. GILMOUR

W. MARGADANT (*Bryophytes & Lichens*)
J. H. PRICE (*Marine Algae*)
and
P. W. RICHARDS (*Bryophytes & Lichens*)

Each entry in the plant lists consists of Johnson's name, followed, in square brackets, by the modern scientific name, or a discussion of this when there is doubt. Some obvious mistakes in Johnson's names have been corrected, and some of his spelling modernised.

At the foot of each page of the translation, before the page number of the continuous pagination, the relevant page number or numbers of the facsimile are given thus: (Facs. p. oo.)

Footnotes signed R. are by Charles E. Raven, those signed L. by Ronald E. Latham, F.R. by Francis Rose, and G. by J. S. L. Gilmour.

For a discussion of the
printer's device, see Introduction p. 5

A Journey
Undertaken for the Discovery of Plants
By Ten Fellows of the Society
Into the County of Kent
In the Year of Our Lord
1629, on July 13

Hampstead Heath
Or an Observation of the Plants growing there
Made in the same year on August 1

Described by the care and pains of
THOMAS JOHNSON

(Facs. p. 29)

A Journey undertaken for the Discovery of Plants

A FEW years ago there grew up among the students of herbalism the truly admirable custom of undertaking, twice or more frequently every year, a three or four days' journey for the purpose of discovering plants. This year the Society decided to travel through part of the County of Kent. The day, July 13, was fixed. Mindful of the date, there assembled early in the morning in St. Paul's Cathedral Master Jonas Styles, William Broad, John Buggs, Leonard Buckner, Job Weale, Robert Larking, Thomas Wallis, two Edward Browns (one of them the apprentice of William Broad) and myself.[1] Then hurrying to the river's bank we hired two boats to take us to Gravesend. But as we left the bank [here follow four lines from Virgil's *Aeneid*, I, 92–5. Canon Raven provided a prose translation, but Mr R. E. Latham has since composed a metrical version, which is printed here, followed by the more nearly contemporary, but less literal, version by Dryden]

Lo, sudden storm clouds from our vision sweep
Daylight and sky. Black night broods on the deep.
Harsh thunder growls. Swift lightnings cleave the gloom.
All powers conspire to threaten instant doom. [R. E. *Latham*

. . . sable night involves the skies;
And Heaven itself is ravished from our eyes.
Loud peals of thunder from the poles ensue;
Then flashing fires the transient light renew;
The face of things a fearful image bears;
And present death in various forms appears. [*J. Dryden*

[1] Five of Johnson's companions (Broad, Buggs, Weale, Buckner, and Larking or Larkin) were among those named by him in the preface to his edition of Gerard's *Herball* (1633) as 'loving friends and fellow Travellers in this study and of the same profession'. Broad contributed commendatory verses to Parkinson's *Paradisus* (1629) and had botanized in Lincolnshire. Buggs had a chequered career: imprisoned as unlicensed, gaining a doctor's degree at Leyden, but forbidden to practise in London, he abandoned medicine for the stage and became one of the Queen of Bohemia's players. For Buggs and Weale see E. Ashworth Underwood, *History of the Worshipful Society of Apothecaries* (1963). For Buckner's career see C. R. B. Barrett, *History of the Apothecaries* (1905), pp. 63, 66, 68. Wallis had been senior apprentice to William Bell of Snow Hill when Johnson was junior. The Brown who was apprenticed to Broad was admitted to the freedom of the Company in 1630. His namesake has not been identified. Styles is probably the Jonas Styles of Cambridge (M.A. 1621) who became rector of Wyham (Lincs.) in 1629 and vicar of Chivelstone (Devon) in 1632: during the Civil War he was 'very ill used and imprisoned insomuch that he was forced to *fly beyond the seas* and was made M.D. at Padua' (J. Walker, *Sufferings of the Clergy* (1714), p. 356). He was restored to his parish in 1660. R. & L.

(Facs. p. 31)

In consequence the boat in which were Buckner, Buggs, Weale and Larking turned aside to Greenwich; and battered by the violence of the weather they went ashore and sought refreshment. We went on to Gravesend without breaking our journey. After dinner we went by the main road to Rochester, leaving a note against the arrival of our comrades at the hostelry at which we decided to spend the night. On the way we found the plants whose names follow:

A lichen plucked from the wall of the inn. [Probably *Xanthoria*, *Lecanora*, etc. spp.]
Cynoglossum maius vulgare. [*Cynoglossum officinale* L.]
Verbena. [*Verbena officinalis* L.]
Irio sive Erismum Dioscor. [*Sisymbrium officinale* (L.) Scop.]
Hyosciamus niger Vulgaris. [*Hyoscyamus niger* L.]
Cicuta. [*Conium maculatum* L.]
Cichoreum sil. [*Cichorium intybus* L.]
Perfoliata. [*Bupleurum rotundifolium* L.]
Ranunculus repens vulgaris. [*Ranunculus repens* L.]
Iacobæa maior & minor. [*Senecio jacobaea* L. and probably *S. erucifolius* L.]
Elaphoboscum erraticum, sive pastinaca siluestris. [*Pastinaca sativa* L.]
Euphrasea vulgaris. [*Euphrasia officinalis* agg., almost certainly *E. nemorosa* (Pers.) Wallr.]
Euphrasea rubra, sive secunda Dodonœi. [*Odontites verna* (Bell.) Dum. (*O. rubra* Gilib.)]
Rubia sylvestris. [*Galium mollugo* L.]
Hedera terrestris. [*Glechoma hederacea* L.]
Ocimoides, Been album Monspelliensium, sive Papaver spumeum. [*Silene vulgaris* (Moench) Garcke (*S. cucubalus* Wibel)]
Medica semine racemoso. [*Trifolium dubium* Sibth.]
Sedum, sive Sempervivum maius vulgare. [*Sempervivum tectorum* L.]
Sedum medium vulg. [Probably *Sedum album* L.]
Sedum tertium Dios. illecebra. [*Sedum acre* L.]
Coronopus Ruellii Cornu cervi alterum vulgi. Lob. [*Coronopus squamatus* (Forsk.) Aschers.]
Hordeum spontaneum spurium Holcus Plinii, Anguillarae. [*Hordeum murinum* L.]
Erigeron, Senecio. [*Senecio vulgaris* L.]
Bursa Pastoris. [*Capsella bursa-pastoris* (L.) Medic.]
Dens Leonis, Taraxacon Rostrum porcinum Caput Monachi. [*Taraxacum officinale* Weber]

(Facs. pp. 31–2)

Solanum hortense. [*Solanum nigrum* L.]
Solanum Lignosum, Dulcamara, vitis silvestris. [*Solanum dulcamara* L.]
Oxalis acetosa vulgaris. [*Rumex acetosa* L.]
Oxalis ovina. [*Rumex acetosella* agg.]
Sonchus aspera vulgi. [*Sonchus asper* (L.) Hill]
Sonchi Lævis species 2 aut 3s. [*Sonchus oleraceus* L., and possibly also *S. arvensis* L.]
Lapathum acutum vulgare. [Fl.K. gives both this and the next as *Rumex conglomeratus* Murr., but *R. crispus* L. seems more likely]
Lapathum acutum minimum Lob. [*Rumex conglomeratus* Murr., and also possibly *R. sanguineus* L. var. *viridis* Sibth.]
Luteola. [*Reseda luteola* L.]
Horminum silvestre. [*Salvia horminioides* Pourr.]
Daucus siluestris vulgaris. [*Daucus carota* L.]
Spondilium, Branca vrsina Germanica. [*Heracleum sphondylium* L.]
Myrrhis siluestris. [Either *Anthriscus sylvestris* (L.) Hoffm. in fruit, or possibly *Chaerophyllum temulentum* L. in flower.]
Lithospermum sive milium solis. [*Lithospermum officinale* L.]
Viorna. Lob. Ἀτραγένη Theophrasti. Clus. Hist. [*Clematis vitalba* L.]
Argentina sive Potentilla. [*Potentilla anserina* L., or possibly *P. argentea* L., which still occurs here]
Sagina spergula. [*Spergula arvensis* L.]
Thlaspi vulgare Mithridaticum, Vaccariæ folio, Lob. [Fl.k. gives *Lepidium campestre* (L.) R.Br., but, from Johnson's other writings, *Thlaspi arvense* L. seems to be intended]
Caucalis silvestris. [Probably *Torilis japonica* (Houtt.) DC.]
Marrubium aquaticum acutum sive Panax Coloni Gerardi. [*Stachys palustris* L.]
Marrubium vulgare. [*Marrubium vulgare* L.]
Ballote, siue Marrubium nigrum. [*Ballota nigra* L.]
Cratæogonon. Lob. [*Melampyrum pratense* L.]
Carduus acaulis Lob. [*Cirsium acaulon* (L.) Scop.]
Carlina siluestris Clusii. [*Carlina vulgaris* L.]
Hieracium maius. [Fl.K. gives *Sonchus arvensis* L., but *Hieracium* sp. (perhaps *H. perpropinquum* (Zahn) Druce) seems more likely]
Hieracium chondrillæ facie. [*Crepis capillaris* (L.) Wallr.]
Hieracium minus præmorsa radice. [*Leontodon autumnalis* L.]
Pseudomelanthium. Nigellastrum, Lychnis segetum [*Agrostemma githago* L.]
Cyanus minor vulgaris. [*Centaurea cyanus* L.]

(Facs. p. 32)

Antirrhinum medium. [*Antirrhinum orontium* L.]
Ladanum segetum Lugd. [*Galeopsis angustifolia* Hoffm.]
Bromos sterilis altera, [*Anisantha* (*Bromus*) sp., possibly a small form of *A. sterilis* (L.) Nevski]
Artemisia vulgaris. [*Artemisia vulgaris* L.]
Papaver rhæas. [*Papaver rhoeas* L.]
Tithimalus helioscopus. [*Euphorbia helioscopia* L.]
Convolvulus minor purpureus & albus. [*Convolvulus arvensis* L.]
Helxine Cissampelos Atriplicis facie, Lob. [*Polygonum convolvulus* L.]
Esula exigua Tragi. [*Euphorbia exigua* L.]
Scabiosa segetum. [*Knautia arvensis* (L.) Coult.]
Gnaphalium vulgare, herba impia Plinii. [*Gnaphalium uliginosum* L., or possibly *Filago germanica* (L.) L., as suggested in Fl.K.]
Elatine Dios. Veronica fœmina Fuchsii. [*Kickxia spuria* (L.) Dum.]
Elatine altera, Lob. [*Kickxia elatine* (L.) Dum.]
Iacea minor purpurea. [*Centaurea nigra* L., probably subsp. *nemoralis* (Jord.) Gugl.]
Cotula fœtida. [Probably *Anthemis cotula* L.]
Anchusa degener facie milii solis. [*Lithospermum arvense* L.]
Rubia sil: purpureis parvis floribus, forte stellaria Lugd. [Probably *Sherardia arvensis* L.]
Perchpier Anglorum Lob. [*Aphanes arvensis* L.]
Saxifraga Anglicana, Alsines minimum genus Daleschampii polygonum selinoides Gerardi. [*Sagina* sp., either *S. apetala* Ard., or *S. procumbens* L.]
Pecten Veneris, Scandix. [*Scandix pecten-veneris* L.]
Rapistrum arvorum flo: luteo & flore albo erucæ foliis. Lob. [*Sinapis arvensis* L., and *Raphanus raphanistrum* L.]
Iacea nigra vulgaris. [*Centaurea nigra* L., probably subsp. *nigra*]
Scabiosa tenuifolia. [*Scabiosa columbaria* L.]
Persicaria, Hydropiper. [*Polygonum hydropiper* L.]
Persicaria mitis maculosa. [*Polygonum persicaria* L., including possibly *P. lapathifolium* L. and *P. nodosum* Pers.]
Bryonia alba, vitis alba. [*Bryonia dioica* Jacq.]
Bryonia nigra. Sigillum B. Mariae. [*Tamus communis* L.]
Polygala flo: rubris. & flo: cæruleis, flos ambervalis Dod. [*Polygala vulgaris* L.; *P. calcarea* F. W. Schultz grows near, but is always blue in Kent]
Lathyris Leguminosa maior Matt. Pisum Perenne vulgo. [*Lathyrus sylvestris* L.; still grows here in plenty, and in Swanscombe Wood (Gerard, 1597)]
Alsine maior glabra. [Fl.K. gives *Myosoton aquaticum* (L.) Moench, but it is

(Facs. pp. 32–3)

unlikely that there were suitable habitats near the highway; perhaps *Stellaria neglecta* Weihe]

Alsine media glabra. [*Stellaria media* (L.) Vill.]

Alsine hirsuta maior. [Probably *Cerastium arvense* L.; it grows today on Johnson's route]

Alsine hirsuta minor. [Fl.K. suggests *Cerastium semidecandrum* L., but it is more likely that it was *C. holosteoides* Fr. or *C. glomeratum* Thuill.]

Centaureum minus flo. albis & rubris. [*Centaurium erythraea* Rafn (*C. umbellatum* auct.)]

Clinopodium vulgare. [*Clinopodium vulgare* L.]

Circæa Lutetiana. Lob. [*Circaea lutetiana* L.]

Onobrychis, sive caput gallinaceum Belgarum. [*Onobrychis viciifolia* Scop.]

Caryophyllata. [*Geum urbanum* L.]

Fragaria. [*Fragaria vesca* L.]

Viola purpurea. [*Viola odorata* L.]

Scorodonia, sive Salvia agrestis. [*Teucrium scorodonia* L.]

Betonica. [*Betonica officinalis* L.]

Crista Galli lutea. [*Rhinanthus minor* L.]

Prunella. [*Prunella vulgaris* L.]

Lycopsis Anglica Lob. [*Echium vulgare* L.]

Helianthemum, a quibusdam Panax chironeum. Cisti species. [*Helianthemum chamaecistus* Mill.]

Nepeta vulgo. Mentha cataria. [*Nepeta cataria* L.]

Origanum onites Matt. Maiorana. Anglica Gerardi, Hyssopus genuina Græcorum. Lob. [*Origanum vulgare* L.]

Pimpinella Saxifraga maior. [*Pimpinella major* (L.) Huds.]

Pimpinella Saxifraga minor, sive Bipennella. [*Pimpinella saxifraga* L.]

At last we crossed the Stone Bridge,[1] an arched and noble structure which spans the river Medway, there a swift and turbid stream, and came to Roffa, called in English Rochester; and not long after we were followed to our inn, which bore the sign of the Bull, by our four Fellows left at Greenwich. When the thunder and rain had abated, they had committed themselves again to the river. But, as the tide was falling, they left the boat at Erith and hurried on by land to Gravesend. They found a few plants not noted by us:

Glycyrrhiza silvestris Gesneri. Glaux vulgaris, Lob. [*Astragalus glycyphyllos* L.; still at Greenhithe. Glaux vulgaris Lob. is a synonym]

Trachelium minus vulgare. [*Campanula glomerata* L.]

[1] Built in 1387 by John Cobham and Robert Knowles 'with the spoils of France' (Camden). See Janet Becker (1930), *Rochester Bridge*. R.

(Facs. p. 33)

Typha maior vulgaris. [*Typha latifolia* L.]

Thalictrum, sive Ruta pratensis Cordi. [*Thalictrum flavum* L.; still grows between Erith and Dartford]

Saponaria. [*Saponaria officinalis* L.]

Then at Gravesend they got word of our departure and followed us swiftly on horseback to the inn. So we were all happy, had supper together, and in our host's garden noted the following plants:

Cardiaca. [*Leonurus cardiaca* L.]

Hyoscyamus luteus. [*Nicotiana* sp.]

Rosmarinus coronarius. [*Rosmarinus officinalis* L.]

'Αργυροκόμη sive gnaphalium Americanum Clusii. [*Anaphalis margaritacea* Benth. & Hook. f.]

Hemerocallis Chalcedonica Lob. [probably *Hemerocallis fulva* (L.) L.]

Chamæcyparissus. [*Santolina chamaecyparissus* L.]

Hyssopus vulgaris sive Arabum. [*Hyssopus officinalis* L.]

Amaranthus purpureus. [*Amaranthus caudatus* L.]

Caryophyllus flo: albidis multiplicibus. [Probably a white-flowered form of *Dianthus caryophyllus* L.]

Delphinium flo: rubris & albis simplicibus. [*Delphinium* sp., probably colour forms of *D. ajacis* L., sens. lat.]

Pulmonaria maculosa. [*Pulmonaria officinalis* L.]

Trifolium odoratum, Lotus odora hortorum Lob. [*Trigonella coerulea* (L.) Ser.]

Costus hortorum. [*Chrysanthemum balsamita* L.]

Salvia. [*Salvia officinalis* L.]

Leucoium, sive Keiri flo: luteo simplici. [*Cheiranthus cheiri* L.]

Beta rubra. [*Beta vulgaris* L.]

Calendula and other similar plants grown everywhere in our gardens. [*Calendula officinalis* L.]

Next morning we made our way to Chatham to see the fleet—in Camden's words, 'the best appointed fleet that ever the Sun saw, ready upon all emergent occasions'.[1] We entered and explored one out of all the fleet, choosing it as a sample of them all, the ship whose name was Prince Royal, and which stands out among all the rest 'As cypress overtops the pliant withies'.[2] This ship has 66 bronze cannon of the larger size; it is so notable in its build, its size, its magnificence, and it so surpasses all expectation that I dare not

[1] William Camden: *Britannia* (1722) (ed. Gibson), *c.* 194. R.

[2] Virgil: *Eclogues*, v, 26. L.

(Facs. pp. 33–4)

describe it, nor if I dared have I the power.[1] After leaving the fleet, a little below the Royal Dock we took to the land, and the plants that follow presented themselves to us:

Limonium copiose florens. [*Limonium vulgare* Mill.]
Glaux vulgaris. [In his *Mercurius* (1634) Johnson uses Glaux vulgaris, Lob. for *Astragalus glycyphyllos* L., and Glaux exigua maritima, Lob. for *Glaux maritima* L.; see also p. 53 above. Presumably he has made a slip here, and is referring to the latter species.]
Spergula marina Daleschamp. Anthylloides Thalii. [Fl.K. does not take this up, but it must be either *Spergularia media* (L.) C. Presl, or *S. marina* (L.) Griseb. See under Rubia marina, p. 59]
Absinthium marinum. Seriphium. [*Artemisia maritima* L.]
Plantago marina. [*Plantago maritima* L.]
Serpentina Matthioli. [*Plantago coronopus* L.]
Caryophyllus marinus, sive gra: marinum. [*Silene maritima* With.]

Then as we climbed the hill we found:

Linum sil: pusillum candidis floribus. [*Linum catharticum* L.]
Acinos vera Anglica Clusii. [*Acinos arvensis* (Lam.) Dandy]
Anonis, siue Ononis. [Probably *Ononis repens* L.]
Serpillum. [*Thymus* sp., probably *T. pulegioides* L., *T. drucei* Ronn. is rare in N. Kent]
Trifolium corniculatum, sive Melilotus coronata Lob. [*Lotus corniculatus* L.]

So we came to Gillingham, dined, and after dinner entered the cemetery and saw:

Ebulus growing in abundance [*Sambucus ebulus* L. No longer there]
Parietaria, [*Parietaria diffusa* Mert. & Koch] &
Filix saxatilis Clusii, both gathered from the Church wall. [Probably *Asplenium adiantum-nigrum* L., but possibly *A. trichomanes* L.]

Nor can I pass over the traditional hospitality given to us and welcomed by us from the vicar of the church, Master Skelton,[2] in whose house moreover we found a lively portrait of a man very famous within our fathers' memory

[1] The *Prince Royal* of 1,200 tons, built in 1610, was by far the largest naval vessel of the time; her complement of cannon, however, was 55, not 66. (See M. Oppenheim: *Administration of the British Navy* (1896), 1, 202.) L.

[2] Master Skelton can scarcely have been vicar of Gillingham: the *Institution Books* at the Public Record Office name as successive vicars James Deyer, presented in 1616 by Edward Hoby, and Charles White (May 1628), Edmund Leigh (June 1629) and Philip Capper (November 1630), all presented by Brasenose College, Oxford. Perhaps Skelton was serving during a vacancy. L.

for his virtue and his zeal for good literature, Sir Edward Hoby, Knight.[1] It bore the inscription 'Things scattered and neglected I assembled'. For he gathered into one collection at great expense and toil the names, ancestral blazonries, and as far as was possible the lifelike portraits of all the Constables (for so we style the prefects of that place) of the Castle of Queenborough, and with his own portrait at the end, all these having been dispersed by the evil of the times and by human negligence.[2]

When we had thanked him, we proceeded by the public highway to the island of Sheppey; and these plants not remarked before were noticed:

Agrifolium. [*Ilex aquifolium* L.]
Taxus. [*Taxus baccata* L.]
Smilax lævis. [*Calystegia sepium* (L.) R.Br.]
Sinapi agreste Apii, sive Laveris folio Lob. [Probably *Sinapis arvensis* L., but possibly *S. alba* L.]
Cornus fœmina. [*Cornus sanguinea* L. (*Thelycrania sanguinea* (L.) Fourr.)]
Bruscus. [*Ruscus aculeatus* L. Still at Newington]
Beta alba. [*Beta vulgaris* L.]
Cyperus rotundus inodorus septentrionalium Lob. [*Scirpus maritimus* L.]
Gra. palustre Cyperoides Lob. [*Carex otrubae* Podp., or, more likely here, *C. divisa* Huds.]
Polypodium. [*Polypodium vulgare* L.]
Calamentha pulegii odore Lob. [*Calamintha nepeta* (L.) Savi; still abundant at Key Street and Bobbing]
Alsine foliis trissaginis. [*Veronica agrestis* L.]
Alsine foliis veronicæ. [*Veronica arvensis* L.]
Gra. tomentosum & acerosum Calamagrostis quorundam Lob. [Fl.K. gives *Calamagrostis epigejos* (L.) Roth, but *Polypogon monspeliensis* (L.) Desf., which grows in Sheppey today, seems as likely]
Gra. tremulum minus vulg. [*Briza media* L.]
Lactuca agrestis odore opii. [*Lactuca virosa* L; still present here]
Pilosella. [*Hieracium pilosella* L.]
Orobanche, rapum genistæ. [*Orobanche rapum-genistae* Thuill.]
Hypericon, everywhere in dry ground. [*Hypericum perforatum* L.]
Teucrium pratense. [*Veronica chamaedrys* L.]
Armeria pratensis, flos cuculi. [*Lychnis flos-cuculi* L.]

[1] Born 1560, Constable of Queenborough 1597, died in the castle there 1617; a friend of Camden (see his *Britannia* (1722), *c.* 195, and *Dictionary of National Biography*). R.

[2] These portraits were commissioned by Hoby about 1593. Sixteen of them were afterwards transferred to Penshurst Place, where two of them can still be seen (see *Gentlemen's Magazine*, **56**, part 1, 56, and *Archaeologia Cantiana*, **58**, 14–25). L.

(Facs. p. 35)

Armeria sil. altera caliculo foliolis fastigiatis cincto &c. Lob. Cariophyllus montan. [*Dianthus armeria* L.]

Lampsana, with many species previously observed. [*Lapsana communis* L.]

We crossed the river, entered the island of Sheppey and finding nothing worthy of mention came to Queenborough. There we put up in two inns. Next morning, while we were about to start our journey, someone came to us and said that the local Prefect (our people call him Mayor) wished to speak with three or four of us. We agreed. Entering his house, we returned his friendly greeting. Then he made an address to us in almost these words. 'The Ancient Kings of this realm decided to grant great and far-reaching privileges to this borough, so that we might the more securely protect this little island of ours from injury. So for me, who am at this time entrusted with the public protection of this place, it is a duty to discover the cause of your journey to this island of ours. It is not that I see anything in you to give me grounds for suspicion, but because it has always been our conviction that it is not consistent with the duties of our Mayors to allow so great a number of men to stay here without our knowing their motives. Tell us therefore in all friendliness the purpose of your travel.'[1] Then briefly answered John Buggs—for this business had been entrusted to him: 'I will expound to you, Sir, the reason for our journey, though it is scarcely worthy of your attention. We are devoted to the study of the science and material resources of medicine. That is why we have come to this place to discover the rare plants that grow in your island. This is the only reason that we have completed this great journey from London as far as this on foot.' Then Master Styles said: 'Another reason besides that stated by my comrade has influenced me—to have the opportunity of seeing a man of such merit as yourself, especially as I know that you are so well versed in seamanship' (for he was a Captain of the Royal Fleet). 'It is especially gratifying to me to become acquainted with so eminent a man.' So by these and suchlike words the Mayor was entirely satisfied. He discoursed briefly about medical and naval affairs: he offered us some excellent beer and, as usual on these occasions, drank our healths: encouraged by his geniality we thanked him, left his village, and betook ourselves to the Castle,

[1] The usual grounds for such investigation were illicit oyster-fishing (oysters were the chief source of Queenborough's wealth) and suspicion of smuggling or even piracy. The ancient charter of the Borough had been replaced in 1626 by a fuller one which granted it the right to hold its own Sessions of the Peace, the mayor being *ex officio* the chief magistrate. It appears from the Queenborough Court of Record Book in the County Archives at Maidstone that the mayor at this date was John Turner. The scantiness of naval records before 1660 makes identification difficult; but he may have been the Captain Turner who was lieutenant to Mansell of the *Lion* in the action at Algiers in 1620 (see J. S. Corbett: *England in the Mediterranean* (1904), I, 127 n.: see also the *Calendar of State Papers Domestic for* 1619–23, p. 100 and that for 1628–9 p. 201). R. & L.

(Facs. pp. 35–6)

once very strongly fortified, the work of King Edward the Third.[1] I do not intend to describe the whole of it, but cannot pass over the dining-hall in which not so long ago the blazonry and names of the nobles of the whole County of Kent were depicted, and at a high level all round the building was written 'To One alone and always'. In the centre was set the royal blazonry of the never sufficiently belauded Queen Elizabeth, and in the same these verses were written in capital letters:

Lilies thy virgin breast, thy royal strain
Lions denote. Live, virgin ! Lion, reign!
Fair image of fair face, and that fair face
Fair image of a mind filled with God's grace,
Let this to me the living image be,
Virgin and queen, of living Deity!
May Judah's lion, Jesse's flower, bless
Thy lions and thy flowers, royal Bess!

Dated 1593
[*Translated by* R. E. *Latham*

We left the Castle on whose summit we collected Ruta muraria [*Asplenium ruta-muraria* L.]; we then hurried to the seashore and our usual task; and on the way were collected:

[*Note.* This list probably refers to the site of the present Sheerness, where shell-sand beaches were almost certainly well developed, as at Grain and Leysdown today. F.R.]

Halimus vulgaris Matt. Portulaca Marina. [*Halimione portulacoides* (L.) Aell.]

Spartum nostras parvum. [Probably *Spartina maritima* (Curt.) Fernald, but possibly *Agropyron pungens* (Pers.) Roem. & Schult.]

Coronopus, stellaria, cornu cervi vulgi. [*Plantago coronopus* L.]

Kali maius, Salicornia. [*Salicornia* sp., probably *S. ramosissima* Woods]

Kali minus. [*Suaeda maritima* (L.) Dum.]

Crithmum Chrysanthemum. [*Inula crithmoides* L.; still common near Queenborough]

Atriplicis marinæ species duæ. [Probably *Atriplex patula* L. and either *A. hastata* L., *A. glabriuscula* Edmonst. or *A. littoralis* L.]

On the shore itself grew:

Brassica marina monospermos Lob. [*Crambe maritima* L. (not *Cakile maritima* Scop. as in Fl.K.)]

[1] Queenborough Castle, built *c.* 1361–6 when the Borough got its first charter, was surveyed in 1650, condemned as useless, sold and wholly demolished. Johnson's description is quoted by Hasted, **6**, 234, who gives a translation of the verses based on a not quite accurate reading. R. & L.

(Facs. pp. 36–7)

Papaver corniculatum flor. Luteo. [*Glaucium flavum* Crantz]
Cochlearia Anglica. [*Cochlearia anglica* L.; still there]
Tithimalus Paralius. [*Euphorbia paralias* L.; still in Sheppey]
Alsines genus pelagicum Clusii & in my opinion, Anthyllis peplios effigie. Lob. [*Honkenya peploides* (L.) Ehrh.; still nearby]
Rubia marina. [Probably *Spergularia marina* (L.) Griseb., or *S. media* (L.) C. Presl]
Eringium marinum. [*Eryngium maritimum* L.; still in E. Sheppey and at Grain]
Kali minimum illecebræ facie. [Possibly a small prostrate form of *Suaeda maritima* (L.) Dum., or, less likely, a prostrate *Salicornia*, e.g. *S. ramosissima* Woods]
Tragum Matthioli. [*Salsola kali* L.]
Bryon lactucæ foliis Plin. Lob. [*Ulva lactuca* L.] See Appendix 1, p. 139.
Lichen marinus. [*Ulva lactuca* L.] See Appendix 1, p. 139.
Quercus marina. [*Fucus vesiculosus* L.] See Appendix 1, p. 144.
Confervæ 2. aut 3 species. [Various filamentous algae]
Algarumq. 2 spe. [Various algae.]

Also collected on the shore—not plants but familiar in medical practice—were Stellamarina [star-fish (*Asterias* sp.)] and Ossasepiae ['bones' of cuttlefish (*Sepia* sp.)].

Then, hiring a barge, we set our course to the Isle of Grain opposite to us, and after leaving the little ship walked five or six miles without seeing a single thing that could give us any pleasure. The road ran along the water's edge. In the heat of the day we were tormented like Tantalus with a misery of thirst in the midst of waters—they were brackish! We were equally afflicted with hunger in that inhuman wilderness where there was no town within reach, no smoke to be seen, no barking of dogs to be heard, none of the usual sights of habitation by which we could arouse our fainting spirits to any breath of hope. At last we overcame our difficulties and thoroughly exhausted arrived at Stoke. When dinner was set and served, the whole company of Fellows, now in a state of fatigue (with the exception of Styles and me), boarded a brewer's dray travelling to Rochester. Thus parted from our comrades, whom we left lolling among the barrels in the wagon and entrusted to the care of the drivers, we set out from Stoke through High Halstow and the hamlet of Cooling and a castle of the same name now partly ruined by age[1], and made

[1] Cooling Castle had belonged to the Cobham family since the thirteenth century. Soon after the accession of James I it was forfeited to the Crown by the attainder and execution of its lord, George Brook, with his brother Henry Lord Cobham; but before 1629 it had been restored to George's infant son Henry (Hasted, 3, 518–20). R. & L.

(Facs. p. 37)

our way to Cliffe; and lest we be taxed with laziness or carelessness we collected many plants, some of them not previously listed, such as:

Vicia sil: Tabernæmontani, Aphacoides. [*Lathyrus pratensis* L.]
Chrysanthemum segetum. [*Chrysanthemum segetum* L.]
Equisetum. [*Equisetum arvense* L.]
Anagallis flo: phœniceo. [*Anagallis arvensis* L.]
Tussilago. [*Tussilago farfara* L.]
Lagopus. [*Trifolium arvense* L.]
Apium palustre in Stoke. [*Apium graveolens* L.]
Sium terrestre. [*Petroselinum segetum* L.; still common at Cliffe]
Cannabis fœmina sativa. [*Cannabis sativa* L.]
Cannabis spuria, Lamium quorundam, Lob. [*Galeopsis tetrahit* L.]
Eupatoreum Cannabinum mas. [*Eupatorium cannabinum* L.]
Eupatoreum Cannabinum fœm. [*Bidens tripartita* L.]
Genista vulgaris. [*Sarothamnus scoparius* (L.) Wimmer]
Genestella infectoria. [*Genista tinctoria* L.; still abundant south of Cliffe]
Lychnis sil: flo: albo. [*Silene alba* (Mill.) E. H. L. Krause]
Papaver spontaneum syl. Lob. [*Papaver somniferum* L.]
Carduus stellatus, Calcitrapa. [*Centaurea calcitrapa* L.]
Carduus Polyacanthus Theophrasti. [*Carduus crispus* L.]
Mellilotus noster vulgare in abundance. [*Melilotus altissima* Thuill.; still plentiful about Cliffe]
Eupatoreum vulgare. [*Agrimonia eupatoria* L.]
Holosteum Matt. Lob. Gramen σπιγονατόκαυλον Thalii. [*Juncus bufonius* L.]
Filix mas. [*Dryopteris filix-mas* (L.) Schott]
Filix fœmina. [*Athyrium filix-femina* (L.) Roth]
Cynoglossum maius vulgare. [*Cynoglossum officinale* L.]
Tragopogon, barba hirci. [*Tragopogon pratensis* L.]
Vlmaria, barba capri, regina prati. [*Filipendula ulmaria* (L.) Maxim.]
Sium Matt. &. Italorum. [Probably *Apium nodiflorum* (L.) Lag.; not taken up in Fl.K.]
Sium erucæfolium, Nasturtium aquaticum vulgi. [*Rorippa nasturtium-aquaticum* (L.) Hayek (*Nasturtium officinale* R.Br.)]
Arundo vallatoria. [*Phragmites communis* Trin.]
Calamintha aquatica Belgarum, Lob. [Fl.K. gives *Mentha arvensis* L., and Gerard's figure appears to be this, but *M. aquatica* L. is also possible.]
Bellis minor, sive Consolida media vulnerariorum, Lob. [*Bellis perennis* L.]

(Facs. pp. 37–8)

Personatia, Bardana, Lappa maior. [*Arctium lappa* L.]
Lysimachia siliquosa maxima hirsuta. [*Epilobium hirsutum* L.]
Lysimachia siliquosa minor hirsuta. [*Epilobium purviflorum* Schreb.]
Coniza maior helenitis. Lob. [*Inula conyza* DC.]
Coniza media. [*Pulicaria dysenterica* (L.) Bernh.]
Sophia Chirurgorum. [*Descurainia sophia* (L.) Webb ex Prantl; still at Cliffe]
Galeopsis vera Dios: Vrtica Herculea Taber. [*Stachys sylvatica* L.]

Then after spending the night at Cliffe, where we heard from our host various rumours about our comrades, next morning we went on our way rejoicing and on the steep hillsides found nothing that we had not noted before. So we left them and on the saltings found the following sea plants not previously noted:

Tripolium maius & minus. [*Aster tripolium* L., probably the rayed and unrayed forms]
Althæa florens. [*Althaea officinalis* L.]
Trifolium fragiferum Clusii. [*Trifolium fragiferum* L.]

In the fields was a heavy crop of Rapum sativum [*Brassica napus* L.] reaped by the sickle. So returning to our usual inn at Gravesend we awaited the coming of our waggoner comrades and gave orders for food to be made ready. While we were dining, news was brought that our comrades were not far off. After a brief delay Thomas Wallis arrived on a horse, which he had hired to ride from Rochester to Dartford, and informed us that he had left Leonard Buckner and Job Weale only a little distance behind him, but that the others would not arrive before evening. While we were awaiting them, the tide ebbed, so that we could not get back that day by way of the river. So we consulted together, all hired horses, and hurried by the king's highway to Dartford. There we dismounted at the hotel bearing the sign of the Bull, and went at once to a place packed with many rare plants called Chalkdale, because stones had once been quarried here for making quick-lime. We found it now decked with grass and many beautiful plants whose names, omitting those already noted, are as follows:

[*Note.* I think I know where Chalkdale may have been at Dartford; it is probable that part of it remains today. F.R.]
Gallium album. [Probably *Galium mollugo* L. subsp. *erectum* Syme, which still occurs at Dartford]
Gallium Luteum. [*Galium verum* L.]
Gentianella autumnalis, not yet in flower. [*Gentianella amarella* (L.) Börner]

(Facs. pp. 38–9)

Viola Calathiana Gerardi, trachelii species. [*Campanula glomerata* L.]

Rheseda, Eruca Cantabrica. [*Reseda lutea* L.]

Pimpinella sanguisorba minor. [*Poterium sanguisorba* L.]

Baccharis Monspelliensum, Coniza maior Matt. [*Inula conyza* DC.]

Campanula minor rotundifolia. [*Campanula rotundifolia* L.]

Centaurium minus luteum. [*Blackstonia perfoliata* (L.) Huds.]

Scabiosa montana minima. [Fl.K. gives *Jasione montana* L., which the name suggests, but probably a slip for *Scabiosa columbaria* L., as Johnson expressly says that the locality is chalky. Johnson may have found *Jasione* later on Dartford Heath, and confused the two records]

Anthillis leguminosa Belgarum. [*Anthyllis vulneraria* L.]

Orchis myodes prima. Lob. [*Ophrys insectifera* L. (*O. muscifera* Huds.); still found not far away]

Serapias candido flo: montana Lob. [A white-flowered *Dactylorchis* (see Gerard, p. 222), probably *D. fuchsii* (Druce) Vermeul.]

Buglossum sil: parvis floribus. [*Anchusa arvensis* (L.) Bieb. (*Lycopsis arvensis* L.); still found about Dartford]

Buglossum echioides luteum. [*Picris echioides* L.]

Caucalis nodoso echinato semine Bauhini. [*Torilis nodosa* (L.) Gaertn.]

Turning to the right [i.e. on Dartford Heath. F.R.] we found:

Erica vulgaris copiosis parvis floribus. [*Calluna vulgaris* (L.) Hull]

Iuniperus cum baccis. [*Juniperus communis* L., in fruit. Once common on all the southern heaths and downs (see Turner, *Herball*, 1562, pp. 2 and 25); probably made scarce by repeated fires]

Evonimus Theoph. [*Euonymus europaeus* L.]

Blattaria flo: albis. [Probably *Verbascum blattaria* L., once frequent in N.W. Kent, though possibly the white form of *V. lychnitis* L., which is still found at Dartford]

So we returned to the town, to the house of Master Wallis,[1] a preacher of divine grace, who welcomed us kindly and refreshed us with a lavish supper. In his paddock were growing:

Ptarmica vulgaris & Bistorta maior. [*Achillea ptarmica* L., and *Polygonum bistorta* L.]

[1] Richard Wallis, presumably a kinsman of Thomas, was vicar of Dartford from 1603 till 1630; the carved wooden pulpit dates from his incumbency (G. A. Tait, *The Church and Vicars of Dartford*). L.

(Facs. p. 39)

On the next morning we got from Saltmarsh:[1]

Anagallis aquatica tertia, Lob. [If the two species of Anagallis in the next list are correctly determined, this may well be *Veronica catenata* Pennell]

Allium sil. tenuifolium. [*Allium vineale* L.]

Here we dined and went on to Erith, leaving on the left the ancient house of Mistress Dakins.[2] We found in the marshes:

Lysimachia purpurea spicata. [*Lythrum salicaria* L.]

Lysimachia galericulata. [*Scutellaria galericulata* L.]

Valeriana sil. maior. [*Valeriana officinalis* L.]

Iuncus floridus, Gladiolus aquaticus, Dod. [*Butomus umbellatus* L.]

Sparganium ramosum. [*Sparganium erectum* L., sens. lat.]

Trifolium paludosum. [*Menyanthes trifoliata* L.]

Alectrolophos, Plin. sive pedicularis rubra. [Fl.K. gives *Pedicularis sylvatica* L., but Johnson does not seem to have distinguished between this and *P. palustris* L., and the latter seems more likely in this habitat]

Acorus nostra palustris cum semine. [*Iris pseudacorus* L.]

Anagallis aquat. Latioribus foliis. flo. cæruleis. [Fl.K. gives *Veronica beccabunga* L., but *V. anagallis-aquatica* L. seems more likely (see *Mercurius*)]

Anagallis aquat. angustioribus foliis, flo. albidis. [Fl.K. gives *V. anagallis-aquatica* L., but probably *V. scutellata* L. if the previous determination is correct]

Myosotis Scorpioides. [*Myosotis scorpioides* L.]

Hydrolapathum magnum. [*Rumex hydrolapathum* Huds.]

Hydrolapathum minus. [Fl.K. gives *Rumex palustris* Sm., but this was not then distinguished from *R. maritimus* L.; the former occurs here now, and the latter was recorded in 1877, so either is possible]

Marrubium aquaticum, sideritis prima. Matt. [*Lycopus europaeus* L.]

Betonica aquatica. [*Scrophularia aquatica* L.]

Plantago aquatica. [*Alisma plantago-aquatica* L.]

Leaving these parts, we noted on the public footpath:

Helenium, siue Enula Campana. [*Inula helenium* L.]

1 Johnson prints the word Saltmarsh in italics, indicating a place name rather than merely a habitat. There is no place of this name known today, but the marshes north of Dartford are still described partly as Dartford Fresh Marshes and partly as Dartford Salt Marshes. F.R.

2 The house of Mistress (or Lady) Dakins cannot readily be identified with any buildings that Johnson would have seen on this trip. It is just possible that the remark may reflect a confused recollection of Lady Dacres who died in 1612 at Chevening—rather a long way to the left of the Dartford–Erith road! (see Hasted, **3**, 110, and *Complete Peerage*, **4**, 12). L.

(Facs. pp. 39–40)

Pes Columbianus, geranium 2. Dios. foliis amplis valde dissectis. [Johnson does not seem to have distinguished between *Geranium dissectum* L. and *G. columbinum* L., and this might refer to either]

Blattaria vulgaris flo. Luteo. [Johnson did not distinguish between *Verbascum blattaria* L. and *V. virgatum* Stokes, and this may have been either]

So we came to Erith, hired a boat, and on our way back saw three ships returned from the East Indies. We boarded one of them setting its course to Blackwall, and Buckner was presented with a big Indian nut, a piece of sugar-cane and an Indian bamboo. We left the ship and crossed the most famous bridge in the British world; then we were told that our Fellows had got back and where they now were. We made haste, found them, and learned from them that after our parting at Stoke they had come in their wagon in great style through Hoo to Rochester and had spent the night in the house of the Mayor where we had stayed before. His name was Allen,[1] and now he welcomed them much more warmly and honoured them with a sumptuous repast. Next morning there came to call upon them Master Larking, Prebendary of Rochester Cathedral, and Master Selby, a gentleman of the same place;[2] to them we are much indebted for their courtesy. Leaving Rochester, on the opposite bank of the Medway at Strood they accepted a kind invitation from Mr Mapeston and Mr Moorland, gentlemen of that place, and from John at Noke de Cliffe to eat lamb prepared in a new fashion. So, after thanking them for their kindness along with Master Pritchard, a student of herbalism who promised to send them Lunaria racemosa [*Botrychium lunaria* (L.) Sw.], they went straight to Gravesend and next day by the main road returned to London. There the Society, once again assembled, decided to explore Hampstead Heath on the first of August; for many plants grow there which are not to be found here.

[1] Anthony Allen was the first mayor of Rochester under the city's new charter of 1629 (see F. F. Smith, *History of Rochester* (1928), p. 52). L.

[2] John Larkin, presumably a kinsman of Robert, was a prebendary of Rochester Cathedral, 1625–54. Henry Selby was vicar of St Margaret's, Rochester, 1627–47 and headmaster of Rochester Grammar School, 1627–31. Both appear in Venn, *Alumni Cantabrigienses*. L.

(Facs. p. 40)

Hampstead Heath

AT the appointed hour in the morning seven of our ten fellows assembled. Buggs, Weale, and Wallis were missing, but their places were taken by John Sotherton,[1] John Marriott, and Thomas Crosse, who joined our fellowship. Undeterred by the lowering sky, we thought it shame that we who had faced greater hazards should shrink from such a trifle. So we left the city and made our way to Kentish Town. From there we had travelled fast but not far when a heavy shower drove us to seek refuge in Highgate. Scarcely had the rain stopped when impatient of delay, we made for the wood [probably Ken Wood]. There, apart from plants observed on our previous outing,[2] which rather than weary my readers with a complete enumeration I purposely omit, we found the following:

Gram. spica Brizae maius Bauh. [Almost certainly *Brachypodium sylvaticum* (Huds.) Beav., though the name refers to *B. pinnatum* (L.) Beauv.]

Tithimalus charachias amygdaloides. [*Euphorbia amygdaloides* L.]

Serratula flo. pur. & albis. [*Serratula tinctoria* L., with purple and white flowers.]

Anagallis flo: luteo. [*Lysimachia nemorum* L.]

Astragalus Sylvaticus Thalii. [*Lathyrus montanus* Bernh. (*L. macrorrhizus* Wimm.)]

Alnus nigra, frangula. [*Frangula alnus* Mill. (*Rhamnus frangula* L.). Fl.M. does not cite this record—only that from the 1632 list]

Sorbus sil: Alpina Lob. Ornus Dod. [*Sorbus aucuparia* L.]

Morsus Diaboli. [*Succisa pratensis* Moench (*Scabiosa succisa* L.)]

Pulmonaria Gallorum Hieracii flore, as I should judge. [*Hieracium murorum* agg., probably *H. exotericum* agg.]

Tormentilla. [*Potentilla erecta* (L.) Räusch. *P. tormentilla* Stokes]

1 John Sotherton became master of the Society of Apothecaries in 1646 (see Kew and Powell, *Thomas Johnson*, p. 28). L.

2 In the phrase 'our previous outing' Johnson is presumably referring to his Kentish journey described in the first part of the *Iter*, though it is perhaps curious that he should consider undesirable an overlap between species recorded in Kent and at Hampstead. This interpretation is borne out by the fact that several species (e.g. *Calluna vulgaris*) which he recorded for Kent, and later for Hampstead, are not mentioned in the *Iter* Hampstead lists. The only other possible interpretation seems to be that Johnson is referring to his visit to Hampstead the previous May (see p. 1), but as far as we know the results of this were not published and, further, he does in fact include some of the 'May plants' (withered in August) in the *Iter* lists. G.

(Facs. p. 41)

Cynocrambe. [*Mercurialis perennis* L.]
Ascyron. [*Hypericum tetrapterum* Fr.]
Populus Lybica. [*Populus tremula* L.]
Betulus, Lob. Carpinus, Matt. [*Carpinus betulus* L.]
Virga aurea Villanovani. [*Solidago virgaurea* L.]
Gnaphalium foliis longioribus. [*Gnaphalium sylvaticum* L.]
Numularia. [*Lysimachia nummularia* L.]
Scrophularia maior. [*Scrophularia nodosa* L.]
Angelica silvestris. [*Angelica sylvestris* L.]
Trifolium acetosum, Oxys Plin. Alleluia sive luiula Offic. [*Oxalis acetosella* L.]

Leaving the wood and coming out on to the Heath proper, we saw:

Erica pumila calyculato vnedonis flore. [*Erica cinerea* L.]
Erica Iuniperifolia altera Lob. [Probably a form of *Erica tetralix* L.]
Vaccinia nigra. [*Vaccinium myrtillus* L.]
Lilium convallium. [*Convallaria majalis* L.]
Genistella aculeata. [*Genista anglica* L.]
Genista aculeata. [This is synonymous with *G. anglica*, and may be an error for *Genista tinctoria* L.]
Lonchitis. [*Blechnum spicant* (L.) Roth]
Muscus clauatus, Lob. Muscus terrestris, Matt. [*Lycopodium clavatum* L.]
Lysimachia galericulata minor, Gratiola latifolia, Ger. [*Scutellaria minor* L.]
Ornithopodium perpusillum. [*Ornithopus perpusillus* L.]
Gram: Sparteum capillaceo folio minimum. [Probably *Nardus stricta* L.]
Clymenon Italorum, Androsæmum quorundam. [*Hypericum androsaemum* L.]
Ros solis. [*Drosera rotundifolia* L.]
Veronica mas vulgaris. Betonica Pauli. [*Veronica officinalis* L., and *V. serpyllifolia* L.]
Veronica pratensis, Dod. [*Veronica serpyllifolia* L.]
Veronica recta minima. Lob. [This name refers to *V. spicata* L., but Fl.M. suggests that a small form of *V. officinalis* was mistaken for *V. spicata*]
Adiantum aureum maius. [*Polytrichum commune* Hedw., perhaps also including *P. formosum* Hedw.] See Appendix III, p. 147.
Adianthum aureum minus. See Appendix III, p. 148.
Muscus pyxadatos alablastriculos imitatus Lob. [*Cladonia pyxidata* (L.) Hoffm. sens. lat.] See Appendix III, p. 148.
Salix humilis bombicifera. [*Salix repens* L.]

(Facs. pp. 41–2)

Iuncus bombicinus. [*Eriophorum angustifolium* Honck.]
Cotiledon aquatica acris. [*Hydrocotyle vulgaris* L.]
Ranunculus rotundifolius, perhaps Apium risus. [*Ranunculus sceleratus* L.]
Ranunculus aquaticus hepaticæ facie, Lob. [*Ranunculus peltatus* Schrank, sens. lat.]
Ranunculus flammeus minor. [*Ranunculus flammula* L.]
Digitalis purpurea. [*Digitalis purpurea* L.]
Aria Theophrasti effigie Alni. [*Sorbus aria* (L.) Crantz]
Sorbus torminalis. [*Sorbus torminalis* (L.) Crantz]
Alsines minimum genus peculiare non discriptum. [This may perhaps refer to *Montia fontana* L., which grew on the Heath. If so, it is the earliest British record]

Leaving the Heath, we rested for a while in the little town of Hampstead. Then, returning to Kentish Town, we had dinner. On our way out and back we observed in meadows, by the wayside, and in hedgerows:

Atriplex sil: sinuata. [*Atriplex hastata* L.]
Atriplex sil: Laciniatis foliis. Pes Anserinus. [Presumably a form of *Chenopodium rubrum* L.; Fl.M. cites 'Johnson, 1629' as the first record, but gives the name used in the 1632 list ('Atriplex sylvestris latifolia sive Pes anserinus')]
Atriplex sil: Polygoni aut Helxcines foliis, Lob. [*Atriplex patula* L.]
Aracus, sive Cracca maior & minor, Lob. [*Vicia sativa* L. and *V. hirsuta* (L.) S. F. Gray]
Bonus Henricus sive tota bona. [*Chenopodium bonus-henricus* L.]
Potomegeiton. [Probably *Potamogeton natans* L., which Johnson records in his 1632 list as P. majus vulgare]
Arum baccis onustum. [*Arum maculatum* L. in fruit]
Geranium Robertianum, Herba Roberti. [*Geranium robertianum* L.]
Geranium secundum Dios. Pes columbinus. [*Geranium molle* L.]
Geranium arvense Tab. Myrrhida Plinii. [*Erodium cicutarium* L.]
Saxifraga Anglicana facie sesili pratensis. [*Silaum silaus* (L.) Schinz & Thell. (*Silaus pratensis* Bess.)]
Holosteum Ruellii, gramen Leucanthemum minus. [*Stellaria graminea* L.]
Barbarea. [*Barbarea vulgaris* R.Br.]

I think it not out of place to note here the names of certain plants found by us in these places on May 1 but now partly or wholly withered. Such are:

Anemone sive ranunculus nemorum flo: albo. [*Anemone nemorosa* L.]
Hyacinthus vulgaris Anglicus. [*Endymion non-scriptus* (L.) Garcke]

(Facs. p. 42)

Primula pratensis. [*Primula veris* L.]
Primula silvarum. [*Primula vulgaris* Huds.]
Cardamine. [*Cardamine pratensis* L.]
Lamium Luteum. [*Galeobdolon luteum* Huds.]
Bugula. [*Ajuga reptans* L.]
Alliaria. [*Alliaria petiolata* (Bieb.) Cavara & Grande]
Caltha palustris. [*Caltha palustris* L.]
Chelidonium minus. [*Ranunculus ficaria* L.]

This brief compilation comprises, with the exception of a few common species, all that was seen in these rambles of ours. I have set it down in order that our friends and others interested in these matters may know how much toil and expense we have undergone to promote the study of botany. So that it may be evident to all that we underwent these exertions not for show but for use, we gathered and preserved any rarities that came our way. But this year's tasks are merely a prelude to others to be tackled in the years that follow, on which may God bestow his blessing. Amen.

THE END

(Facs. pp. 42–3)

Thomas Johnson's
Descriptio Itineris, 1632
in facsimile

REPRODUCED FROM THE COPY IN
THE BRITISH MUSEUM, LONDON
BY PERMISSION OF
THE TRUSTEES

Descriptio

Itineris Plantarum

*Investigationis ergô sus-
cepti, in Agrum Cantianum
Anno Dom.* 1632.
Et

Enumeratio Plantarum *in Eri-
ceto Hampstediano locisq;
vicinis Crescentium.*

*Sibi ne invideatur ipsi ignavi rectè cavent.
Plaut. in Bacch.*

Excudebat, *Tho. Cotes.* 1632.

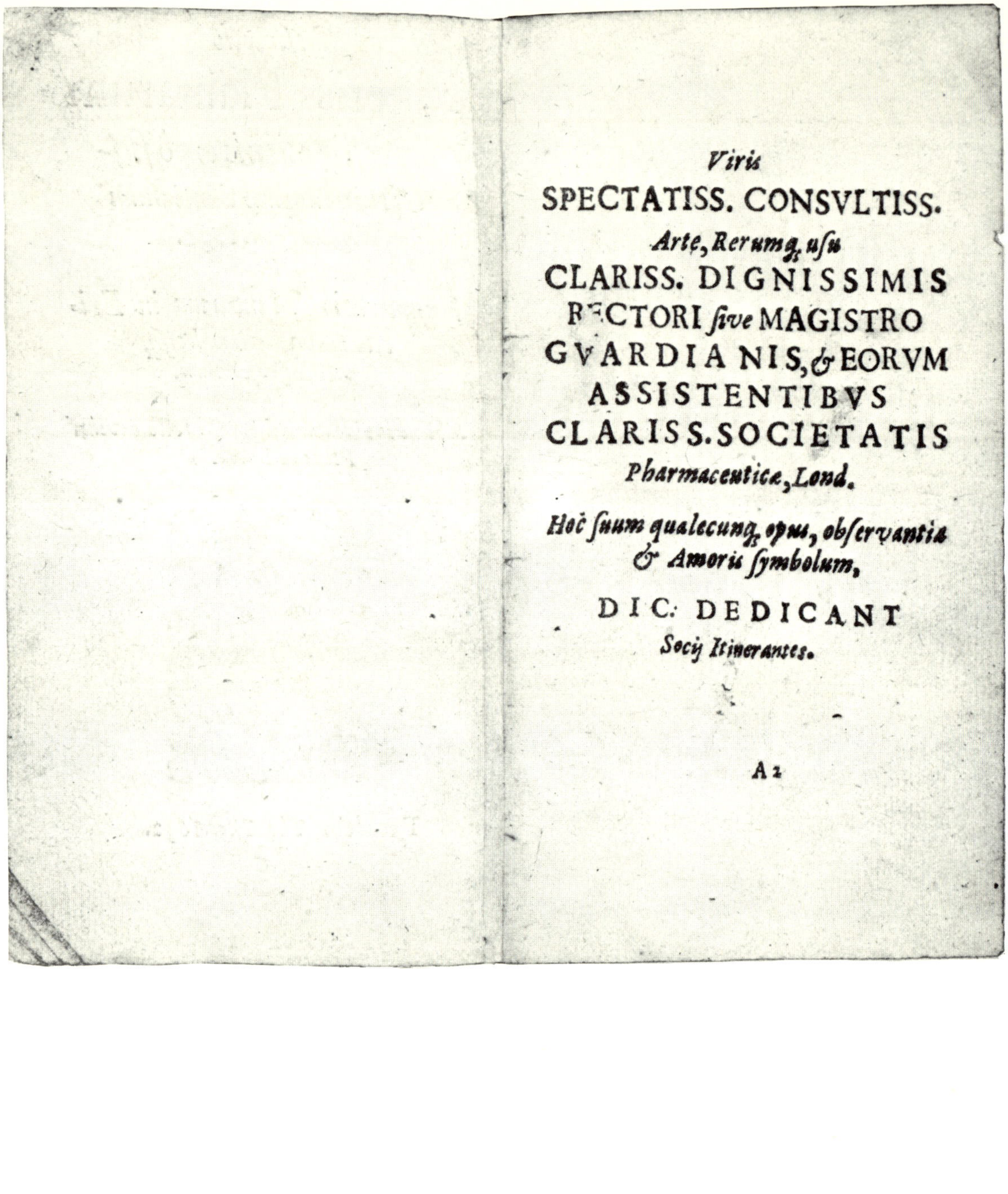

Viris

SPECTATISS. CONSVLTISS.

Artę, Rerumq; usu

CLARISS. DIGNISSIMIS

RECTORI *sive* MAGISTRO

GVARDIANIS, *&* EORVM

ASSISTENTIBVS

CLARISS. SOCIETATIS

Pharmaceuticæ, Lond.

Hoc suum qualecunq; opus, observantiæ & Amoris symbolum,

DIC. DEDICANT

Socij Itinerantes.

A2

Amicis Lectoribus.

Ante biennium Plantarum investigationis ergô (secundum consuetudinem nostram) susceptum Iter, festinante calamo descriptum, typis, amicorum gratiâ, mandavi. Sed mirum quam variam & fortunam & censuram (quamvis private paucorum gratia excussum fuit) expertum est. Nonnulli enim non solum laborem nostrum ut inanem & supervacaneum, sed omnem exactiorem plantarum scientiam, ut inutilem deriserunt, satis esse judicantes illas nomine, & lectione tenus noscere. Sed certissimum est veteres qui medicinam fecerunt, non adeo socordes, & ignaros fuisse; hocq; (omissis Galeni & aliorum tam veterum quam neotericorum testimoniis) ex Oribasio probare facile est: Verba ejus hæc, sunt; Ἡ γνῶσις τῶν ἁπλῶν φαρμάκων, καὶ τῶν ἐν αὐτοῖς δυνάμεων οὕτως ἀναγκαιοτάτη καθέστηκεν, ὥς μὴ δ' οἷόν τε εἶναι χωρὶς ταύτης ἰατρεύειν ὀρθῶς: *&c. id est. Simplicium medica-*

mentorum & eorum facultatum cognitio sic summè necessaria est, ut absq; hac rectè medicinam facere non detur. Non simpliciter ἀναγκαῖαν, ἀλλὰ ἀναγκαιοτάτην τὴν γνῶσιν τῶν ἁπλῶν φαρμάκων esse dixit. Sed quod ad ignavos illos Pharmacopæos nostrûm insulsos derisores attinet, paucis eorum summam ignavia, aut arrogantia ortam ignorantiam patefaciam & depingam. Quotidie fere in foro herbario eorum unus, aut alter (magno ægrotorum dispendio) fœminis rhizotomis se deridendum præbet. Illæ enim imperitos optime notos habentes, illis quicquid pro quolibet audacter obtrudunt. Et eorum non pauci ab illis sumunt Cotyledonem aquaticam acrem pro Vmbelico veneris; Sedum minimum Diosc. 3. pro Sedo minore (i.) vermiculari, Offic. Polygalam flore cæruleo pro Gratiola, Anagallidem flo. phœniceo, pro utraq; pimpinella, tam sanguisorba, quam saxifraga: & minore dispendio, sed non minus crasso errore Sambucum pro Ebulo, Bonum Henricum pro Mercuriali, filicem marem pro filice florida (i.) osmunda regali; non pauca alia enumerare

merare

merare possum sed hæc satis, superq; illos crassæ ignorantiæ reos arguere sufficient. Nonne miseranda ægrotorum talium Medicorum, & Pharmacopæorum opera vtentium conditio? dum medicus Pharmacopæo, ille vetulæ avaræ & sordidæ (quæ aliquid ipsi imponere & ausit, & possit) fidem habet; sic enim aliquando fit, vt ex ignaræ & vafræ fœminæ cognitione herbaria eorum salus pendeat. Desinant ergo in posterum aliorum publici boni gratia susceptos labores vt vanos & invtiles deridere, & calumniari, cum ipsi non solum opprobrio, sed etiam graviore pœna digni sunt. Et hæc sunt ea, quæ pro nobis respondere habui, vt voluptatem quam maledicendo habuerunt, eam male audiendo amittant. Sed posthac speramus harum calumniarum loco illam ἔριδα concitare, de qua extat hoc Hesiodi Elogium, Ἀγαθὴ δ' ἔρις ἥδε βροτοῖσι; quod & vestrûm & Reipublicæ causa ex animo optat,

THOMAS IOHNSON.

A4

Descriptio Itineris Plantarum investigationis ergô suscepti in Agrum Cantianum.

1632.

HOc anno Societate Pharmacopæorum Londinensium solennem plantas inveniendi gratiâ perambulationem peragente; hujus nostræ societatis Guardianus *Thomas Hickes* nos (qui nonnullis præteritis annis soliti fuimus plantarum solum natale lustrandi causa triduum aut quatriduum suscipere iter) hortatus est: ut solito more aliquot dies insumeremus, & sumptu se nos præire, & labore associare promisit. Volentibus persuadere haud erat difficile, statim Annuimus, & ei gratias reddimus, quod nos tanto honore dignaretur. Igitur Augusti primo ad hoc parati *Gulielmus Broad, Leonardus Buckner, Robertus Lorkin, Iacobus Clarke, Thomas*

Iohnſon, convenerunt & ingreſſi ſunt domũ *Thomæ Hickes* Guardiani noſtri. Hinc (ſumpto prius ejus munificentia jentaculo, & non parum viatici) ad Myoparonẽ tendimus, quo conſcenſo & velis ventodatis, Londinum reliquimus, & decem horarum ſpatio, Sexaginta milliaria noſtratia emenſi ſumus, ita ut incipiente primum nocte ad cretaceos Tenet inſulæ clivos, & Margetæ ſinum appulſi ſumus, vbi defixæ & compactiles trabes, injectaque ſaxa molem ſeu Pilam ad commodiorem navium ſtationem faciunt; hic nave egreſſi ad hoſpitium tendimus, vbi omnia apta & parata habuimus, vt etiam hoſpitem noſtrum ſtudioſiſſimũ (*Richardus Pollard* ei nom en eſt) Ille dum in Inſula manſimus nunquam a nobis diſceſſit, nec in diſceſſu aliorum more in nos prædatus eſt. Hic paululum refecti cubitum nos conferimus.

Poſtera Luce egreſſi ad munimentum clivoſo promontorio impoſitum, & natura magis quam arte munitum, in ipſo littore, clivoſiſque collibus collegimus plantas, quarum nomina ſequuntur.

Bryon Lactucæ folijs, Lob. Lactuca marina, Taber.
Alga membranacea ceranoides, Bauhini.

Fucus

Fucus longiſſimo, latiſſimo, craſſoque folio, Bauh. prod. forté Phaſganoides ejuſdem, nomen enim rei bene convenit, ut etiam, Alga latifolia coriacea, ejuſdem, & fere affirmare auſim, ſi non eædem, ejuſdem ſpeciei varietates ſint. Vide figuram hujus in fine lib. notatam, fig. 1. & varietat. polyſchid, fig. 2.
FVci ejuſdem varietas polyſchides.
Alga membranacea ceranoides, Bauh.
Fucus marinus quartus, Dod.
Fucus ſpongioſus ramoſus.
Fucus ſpongioſus nodoſus. Cujus figura notatur, fig. 3.
Fucus ferulaceus, Lobelij.
Quercus marina vulgaris folijs non ſerratis.
Quercus marina folijs ſerratis.
Corallina ramoſa parva, & pennata longior.
Gramen caninum geniculatum marinum, Lob.
Gramen parvum marinum ſpica loliacea.
Cornu cervinum hirſutum, Coronopus, ſive Herba ſtella, Mat. Lob.
Tragon improbus, Matth.
Papaver corniculatum flore luteo.
Papaver erraticum minus.
Bellis minor vulgaris.
Alſine corniculata, Cluſij.

Anthyllis

Anthyllis leguminosa Belgarum, Lob. *Lagopodium*, Tab.

Trifolium fragiferum, Clus. *floribus dilutè rubentibus*.

Trifolium luteum minimum, Lob. *sive medica sem. racemoso.*

Trifolium luteum majus lupulinum, sive lupulus silvaticus, Thal.

Trifolium corniculatum, Dod. *Melilotus coronata*, Lob. *minor glabra.*

Pimpinella saxifraga major, Dod. Lob. *Tragoselinum majus*, Tab.

Pimpinella saxifraga minor, sive Bipinella Lob.

Onopyxos Dodonai, Lugd.

Carduus stellaris, Dod. *seu Calcitrapa, Ejusd.* & Lob.

Rheseda vulgaris, eruca perigrina, Italica vel cantabrica, Lob.

Sinapi sativum vulgare.

Sinapi alterum siliquâ falcatâ, sive sinapi alterum sativum, Lob.

Vrtica non mordax vulgaris purpurea.

Vrtica silvestris asperior, sive vrtica fæmina.

Vrtica minor acrior.

Cotula fœtida vulgaris.

Cotula fœtida flore elegante multiplice.

Gallium luteum.

Spumeum papaver, Lob. *vulgò behen album*;

bum; *Herba articularis*, Tab.

Horminum sylvestre, sive Oculus Christi, est Hormini silvestris, 4. *species* 5. Clus. Hist.

Bursa pastoris major & minor.

Dens leonis, caput monachi, rostrum porcinum, Taraxacon major & minor.

Trifolium pratense album.

Scabiosa minor, sive columbaria.

Gramen Typhinum minus.

Caucalis nodoso echinato semine, Bauh.

Rubia sylvestris.

Acanthium sylvestre, sive Onopirdon.

Lycopsis Anglica, Lob. *Echij altera spec.* Dod.

Atriplex marina repens, Lob.

Atriplex silvestris polygoni, sive helxines folijs.

Helxine, sive parietaria, copiose in sepibus.

Elephoboscum erraticum, Tab. *sive pastinaca sylvestris*, Dod.

Irio, sive Erysimum vulgare, Lob. *Verbena fam.* Trag.

Anonis, sive ononis non spinosa.

Lapathum acutum vulgare.

Alsine media glabra.

Arthemisia vulgaris.

Euphrasia rubra sive secunda, Dod. *Odontites*, Tab. *Sideritis pratensis rubra.* Lugd.

Lugd. *Crataogonon Euphrosine*, Ger: *Luteola*, Lob. *Pseudostruthium*, Matt. *Antirrhinon*, Trag. *catanance*, Tab.

Daucus silvestris vulgaris.

Rapistrum arvorum, Lob.

Cicuta vulgaris.

Cicutaria fatua, Lob. *Petroselini vitium*, Tragi.

Ballote, sive marrubium nigrum.

Sedum minimum, sive. 3. Dios. Illecebra, Dod.

Erigeron, sive senecio minor vulgaris.

Erigeron tomentosum alterum, Lob.

Hieracium montanum saxatile, Columna.

Tum ad hospitium revertentes expectamus Simonem Rose qui eo Loci practicam medicinam exercet, ut eum per insulam ducis loco ad plantarum investigationem haberemus; Ille parvâ interpositâ morâ adest, nosque præit, sed non vulgari tramite, ad Nash: in itinere collectæ sunt plantæ sequentes.

Telephium vulgare, Fabaria Matth. Faba inversa, Lob.

Hordeum spontaneum spurium, Holcus Pliny, Anguil.

Iacobæa vulgaris major. Senecio major, Matth.

Iacobæa minor folijs magis dissectis.

Gnapha-

Gnaphalium vulgare, filago, centunculus.

Anagallis flore phæniceo.

Ranunculus vulgaris pratensis repens.

Ranunculus vulgaris pratensis surrectis cauliculis.

Sophia Chirurgorum, Lob. Seriphium Germanicum, Tragi, Thalictrum, Tab.

Gramen cristatum, Bauh.

Lapathum acutum minimum, Lob.

Trifolium pratense purpureum.

Tithymalus helioscopius.

Cyanus minor vulgaris, Baptisecula, Trag.

Iacea nigra vulgaris.

Iacea major, Lob. flos Cyanoides, Dod.

Helxine cissampelos altera atriplicis effigie, Lob. volubilis nigra, Tab. Orobanche, Tragi.

Chrysanthemum segetum, Lob. Clus. Bellis major lutea, Trag.

Elatine Dioscor, sive veronica fœmina, Fuch. & Dod.

Elatine, Matth. Elatine altera, Dod. Lob.

Verbena.

Cuscuta, Matth. Androsace, Tragi.

Mentha Cattaria, Lob. Nepeta vulgaris, Trag.

Asperula cærulea arvensis.

Myosotis

Myoſotis ſcorpoides, Lob. Euphraſia cærulea Tragi, & Tab.

Ladanum ſegetum Plinij, & Tetrabit anguſtifolium alijs, Lugd. Alyſſum Galeni flore purpureo, Tab.

Coniza media Matth. Calamintha 3. genus, Fuch.

Coniza major Matth. Baccharis monſpelienſium, Lob.

Millefolium vulgare floribus albis, & flo. Carneis.

Scabioſa major ſatorum.

Scabioſa floribus albis.

Pſeudomelanthium, nigellaſtrum, lychnis ſegetum.

Marrubium aquaticum acutum, Gerardi, & panax Coloni ejuſdem, Stachys paluſtris, Geſ. Sideritis gravis odoris, Thal.

Calamentha vulgaris, Offic.

Lampſana, Lob. Sonchus ſylvaticus, Tab. & Ger.

Hieracium minus præmorſa radice, Lob. Hieracium leporinum, Ger.

Hieracium intubaceum, Tab.

Lichen, hepatica vulgaris.

Virga paſtoris, Matth. Lob. Dipſacus tertius Dod.

Caryophyllata vulgaris.

Plantago major vulgaris.

Plantago minor quinquenervia, Lob.

Plantago

Plantago pannicula ſparſa, ſive ſcoparia, Lob.

Serpillum.

Eſula exigua, Trag. Tithymalus leptophyllos, Matth.

Alſine ſive Hippia major, Lob.

Alſine folijs Triſſaginis, Lob. Morſus Gallinæ, 3. Trag.

Alſine folijs veronicæ, Tab. Elatine polyſchides, Lugd.

Peirchpeir Anglorum, Lob. Alchimilla montana, Col.

Morſus gallinæ folio hederulæ alter, Lob. Ballote criſpa minor, Lugd.

Vrtica heraclia, Tragi. Herculea, Tab, Galiopſis Dioſcoridis, Cluſ.

Lychnis ſilveſtris flo. albo.

Eupatoreũ vulgare, ſive Agrimonia, Offic.

Caucalis ſemine aſpero floſculis ſubrubentibus, Bauh.

Myrrhis ſilveſtris, ſive Cerefolium ſilveſtre, Tab.

Prunella vulgaris, Tragi. conſolida minor, Matth.

Periclymenum non perfoliatum, Caprifolium, Offic.

Viburnum, Matth. viurna vulgi gallorum & Ruellij, Lob.

Geranium Rupertianum, ſive herba Roberti.

Geranium

Geranium alterum Dioscor, sive Columbinum. Tab. pes columbinus, Dod, & ejus altera species folijs majoribus & magis dissectis.

Evonymus Theophrasti, Lob. Dod. Carpinus Theophrasti, Trag.

Hedera vulgaris.

Pecten Veneris, Matth. Scandix, Dod.

Clinopodium vulgare, Matth. Origanum, 4. Tragi.

Sonchus arborescens, Tab. Ger.

Gramen spica Brizæ majus, Bauh.

Sic tandem ad nash pervenimus, & a loci Domino benigne excipimur; In ejus horto crescebant.

Draco hortensis, sive Draco herba, Dod.

Raphanus rusticanus.

Antirrhinum majus purpureum.

Cataputia minor, Lob. Lathyris, Matth. Trag. Dod.

Sisarum, sive siser, Matth.

Levisticum vulgare, Dod. Ligusticum, Tab.

Malua hortensis rosea.

Hepatica nobilis, Trag. Herba trinitatis Matth.

Rha rotundifolium, sive Hippolapathum, rotundifolium, Lob.

Sedum serratum flore rubente maculato, Par-

Parkinsoni.

Auricula vrsi, Matth. Lunaria arthritica Gesn. Alysma, seu Damasonium, Dioscoridis, Col.

Helleborus niger verus flo albo.

Flos solis major, sive Chrysanthemum Peruvianum.

Melissa vulgaris.

Napellus verus cæruleus, Ad. Lob. Aconitum cæruleum Gesneri.

Asarum.

Hemerocallis Chalcedonica.

Martagon. sive Lilium cruentum.

Pulmonaria maculosa.

Caryophyllus pratensis flore pleno, Bauh. Odontitis flore pleno. Cam. Clus.

Iacea sive Flos trinitatis, Matth. viola tricolor, Dod. Clus.

Hisce observatis, ad hospitium reversi, & pransi Ductoris suasu ad Queakes cursum dirigimus, & curiosa sedulitate omnia lustrantes, sequentes non prius in itinere collectas plantas invenimus; fuerunt.

Sphondylium hirsutum vulgare; Branca vrsina, Trag. Acanthus, vulgaris, Fuch.

Scrophularia major, Lob. Ocimastrum alterum, Trag.

Hieracium folijs & facie Chondrillæ, Lob.

Lau-

Laureola, Dod. Daphnoides, Lob. Tab. Thymelea, Trag.

Hypericon vulgare.

Aſcyron, ſive Hypericon in dumetis, 2. Trag.

Centaurium minus vulgare.

Xyris. ſpatula fœtida, Offic.

Oxalis. ſive acetoſa vulgaris.

Polygonum mas vulgare, Lob.

Lithoſpermum, ſive milium ſolis, Offic.

Carlina minor flore purp. Cluſ. carduus acaulis, Lob.

Geranium arvenſe, Tab. Cicutæ folio, Bauh.

Viorna vulgi, Lob. Ἀτραγένη *Theophraſti, Cluſ. Clematis 3. Matth.*

Ptarmica vulgaris, Lob. Pyrethrum ſilveſtre, Dod. Tanacetum album, Trag.

Solanum lignoſum, ſive Dulcamara, Dod. Vitis ſyl. Matth.

Lupulus ſalictarius.

Sanicula vulgaris, Diapenſia, Matth.

Phalaris, ſemen canarienſe vulgò Belgis & Anglis.

Bryonia nigra, Sigillum Beatæ Mariæ Offic. Tamus, Dod.

Bonus Henricus, Trag. Matth. Rumicis 3. Genus, Fuch.

Cynocrambe, ſive mercurialis ſylveſtris, mas & fœm.

Fra-

Fragaria vulgaris.

Siſon, Cordi, Petroſelinum Macedonicũ, Fuch. Dod. Amomum Germanicum, Tragi.

Acer majus, Lob. Platanus Scotica, Cardani.

Acer minus vulgare.

Lychnis ſilveſtris flore rubello.

Rhamnus Catharticus, ſive ſpina infectoria, Matth, & Lob.

Filipendula, Saxifraga rubra, Tab. Oenanthe Lob. Fuch.

Praſſium, Marrubium album.

Chelidonium majus.

Lyſimachia ſiliquoſa media.

Argemone capitulo torulis canulato, ſive Argemone Lacunæ, Lob.

Argemone capitulo longiore, Ejuſdem.

Polygonum ſelinoides, Ger. ſive Knawel Germanorum; an Vermiculata nova planta montana, Col?

Acinos parva flore cæruleo, ſive Acinos Anglica, Cluſij.

Pedicularis pratenſis Lutea, ſive Criſta Galli herbariorum, Lob.

Rubus ſaxatilis, Cluſij.

Barbarea, ſive pſeudobunias, Lob.

Sonchus aſper vulgaris.

Sonchus Lævis tenerior, Lob. folijs minus diſſectis.

Sonchus

18

Sonchus Lævis vulgaris folijs laciniosis, Ejusd.

Hipposelinum vulgare, Olusatrum, Petroselinum Alexandrinum, Trag.

Blattaria flo. luteis vulgaris.

Alsine corniculata, Clusij.

Absinthium marinum, sive Seriphium.

Alsine cruciata marina; vel Alsine pelagicum, Clusij. Anthillis prior lentifolia peplios effigie maritima, Lob. Ad.

Kali majus, sive salicornia.

Kali minus.

Soldanella, Κράμβη θαλασσία, *Dios.*

Brassica marina, Monospermos, Lob.

Erucamarina, sinapi marinum Ægyptium, Alpini.

Serpentina major, Matthioli.

Eryngium marinum vulgare.

Glaux exigua maritima.

Spergula marina, Lugd. Forté Anthylloides, Thal. ad salinas Saxonicas, & Alsine maritima Neapolitana, Col.

> *Hæc crescebant in parvo sinu dicto, vulgò, West gate Bay.*

Cyperus rotundus inodorus septentrionalium, Lob.

Paludapium, sive Apium, Offic. magna copia

pia

19

pia ubique in uliginosis, & salsis.

Beta silvestris spontanea maritima, Lob.

Tota die in hisce inveniendis consumpta, post solis occasum defessi domum nos conferimus. Dein cibo, somnoque refecti, sequente mane Margata post tergum relicta, Sandwich versus tendimus, & antequam ad Maris littus venimus, collectæ fuerunt plantæ sequentes.

Euphrasia vulgaris.

Sideritis alsines trixaginis folijs, Bauh.

Oenanthe Angustifolia, Lob.

Cicutaria palustris, Lob. Phellandrium Plinij. Dod.

Sagittaria aquatica, Plinij, Major Matth. Dod. Phleos mas latifol, Lugd.

Hydrolapathum minus, Lob.

Anagallis aquatica major folijs acutioribus, floribus albidis.

Polygonon fæmina semine vidua, Lob.

Equisetum, 1. Matth. Hippuris major, Dod.

Arundo vulgaris vallatoria, Lob. palustris, Matth.

Hydropiper, persicaria acris;

Persicaria mitis maculosa.

Iacea Floribus albis.

Gram. spica Tritici mutici, Bauh.

Cichoreum sylvestre flore cæruleo.

Tragopogon vulgare luteo flore.

Cyno-

Cynogloſſum majus.
Bardana, Lappa major.
Gallium album, Tab. paluſtre, Dod.
Chamæpitys, ſive Iva arthritica, Offic.
Carlina ſilveſtris Dod. vulgaris, Cluſ. Atractylis, ſive Cartamus ſyl. Trag. carduus vulgaris Matth. Hieracantha, Tab.
Solanum lethale, Dod. mandragora Theophraſti, Bella donna; Italorum.
Eupatorium cannabinum mas. Herba St. Kunigundis, Trag. Eupatorium vulgare, Matth. Dod. Eupatorium Avicennæ vulgò creditum.
Lyſimachia ſiliquoſa maxima hirſuta.
Lagopus, ſive pes leporis.
Ladanum ſegetum, Lugd. flore albo.

Caryophyllus marinus, ſive gramen marinum.
Iuncus major durior.
Gramen ſparteum 2. Schœnanthinū, Tab. Spartum noſtras, Lob. Spartum. 3. Cluſ.
Lychnis marina Anglica, Lob.
} *Ad maris littus.*

Tum demum trajecto amni, é Tenet diſcedentes, Sandwich venimus, ingreſſoque hoſpitio, illic paululum moramur. Dein ad maris littus Sandowne Caſtrum ..rſus duo amandantur, dum reliqui oppidum

dum luſtrare ſe accingunt; qui ductu D. Sparkes pædagogi, muros, & munimenta jam partim vetuſtate lapſa circumambulant, & hortum Gaſpari Nirenij Belgæ, ingrediuntur; vt etiam Officinam Pharmaceuticam Caroli Anati (cui poſtea Cantuariæ obvij facti ſumnes) quo in loco rem memoriæ dignam viderunt, ſpolium (ut ſic loquar) Serpentis quindecim pedes longi, & plus quam brachialis craſſitudinis. Quantum conjecturâ aſſequi poſſim fuit Serpens marinus, captus enim erat a duobus viris, inter arenoſos tumulos ad maris littus, capite prius glandibus minoribus machinâ ignevoma emiſſis ſpoliatus. Ex cuniculis, qui illic magna ſunt copia victum querebat, namque, ex ejus ſtomacho eorum vnus et alter extracti fuerunt. Sed hi, beſtiam, ut dixi, vita ſpoliatam ad noſtrûm amiciſſimū Carolum Anatum detulerunt, & eam, accepto præmio ei dederunt, qui carne abjectâ, pellem fæno farctam ſecum in rei memoriam adhuc ſervat. Ex horto Nirenij, Maris Littore, viciniſque locis habuimus ſequentia.

Malva vulgaris floribus albis.
Anagallis aquatica. 3. Lobel.
Cotyledon aquatica ſeu acris ſeptentrionalium, Lob.

B Be-

Buglossum silvestre, Dod.
Tithymalus paralius.
Ossa sepia.
Stella marina.
Solen fœmina.
Alcyonij dua species.
Pectunculus, Rond.
Echinus marinus sine aculeis minor, Besleri.
Echinus marinus, spatagus, Rond. sed adeo tener & fragilis fuit, ut vix manibus tractari potuit.
Umbilici parvi, Boussueti.
Concha longa altera, Ejusdem.
Lepades varia magnitudinis.
} *Marina.*

Myrrhis, Matth. Dod. Cerefolium Hispanicum, Tab.
Glycyrhiza vulgaris.
Mercurialis mas, & fœmina.
Valeriana græca, Dod.
Ageratum, Diosc. Eupatoreum Mesue, vulgò.
Bistorta, Offic. serpentaria mas, Fuch.
Caryophyllata montana, Matt.
Sambucus rosea.
} *Ex horto Nirenij.*

Nocte ingruente reversi ad cœnam discubuimus non tam multiplicibus ferculis,

lis, quam innoxijs jocis refertam, sicque præterriti itineris tædium levati ad quietem secessimus. Postera luce locario, sumptibusque reliquis stabulario persolutis, Sandwich reliquimus. Sicque destinatum iter Cantuariam versus aggressi a nobis colliguntur.

Morsus Ranæ, Dod. Lob.
Iuncus floridus, Gladiolus aquaticus, seu palustris Cordi.
Potamogeiton angustifolium.
Ranunculus tricophyllon aquaticus medioluteus, Col, sive millefolium maratriphyllon tertium, &c. Lob.
Viola aquatilis, Dod. Palustris, Ger.
Ranunculus aquaticus hepaticæ facie, Lob. Polyanthemum aquat. Dod.
Raphanus aquaticus, Tab.
Sium umbellatum repens. Matthioli & Italorum, ut puto.
Sium Crateva erucæfolium; Lob. sive Nasturtium aquatic. Offic.
Armeria pratensis, Lob. Caryophyllus pratensis, Tab. Flos cuculi pratensis, Trag.
Lens palustris.
Sparganium ramosum, Platanaria, Dod.
Typha major.
Thalictrum, sive Thalietrum, Ruta pratensis, Cordi.

B 2 *Oenanthe*

Oenanthe aquatica minor, Iuncus odoratus aquatilis, Dod.
Lysimachia purpurea spicata Lob. Solidaginis saracenica alia species, Trag.
Buglossum echioïdes luteum, Lob.
Calamentha aquatica, Belgarum.
Gramen palustre Cyperoides, Lob.
Dipsacus vulgaris.
Carduus lanceolatus, Tab.
Carduus sylvestris primus, Dod.
Carduus sylvestris 3. Ejusd. sive Polyacantha Theophrasti, Lob.
Marrubium aquaticum, Sideritis prima, Matth.
Betonica aquatica.
Filix mas.
Filix femina.
Adianthum album Plinij, Lob.
Genista vulgaris.
Sagina spergula Lob. Anthylloides, Thal. Alsine terrestris altera τριχόφυλλος, *Col.*
Spergula flore rubro; an Alsine spergulæ facie minor, Bauh. Saxifraga Antiquorum. Adver.
Ranunculus arvorum semine echinato.
Lychnis silvestris parva.
Cracca major, Aracus Lob.
Festuca altera, Dod. Lolium. 1. Trag. Bromus sterilis altera, Lob.
Thlaspi vulgare, sive vaccariæ folio, Lob.

Loli-

Lolium, sive Triticum temulentum, Lob.
Carduus vulgatissimus radice repente, Κεάνωθος *Theophrasti, Col.*
Gramen milliaceum.
Tussilago farfara, ungula Caballina.
Symphytum majus, sive Consolida major vulgaris.
Petasites, sive Tussilago major, Matth.
Hypericum minus repens.
Veronica minor serpillifolia, Lob.
Aracus, Tab. Galega silvestris, Dod.
Carduus mariæ, Leucographis Plinij.
Linum silvestre perpusillum flo. albis non descriptum.
Vicia, Tab. Legumen terræ glandibus simile, Dod. Aphacoides.
Caltha palustris, Tussilago altera, Matth
Acorus nostras palustris, Lob. Iris palustris lutea, Tab. Pseudoiris, Dod.
Iuncus capitulis sparsis.
Trifolium corniculatum majus hirsutum
Anchusa degener facie milij solis, Lob. Anchusa arvensis minor, Tab. Lithospermum silvestre, Trag.
Anonis sive Ononis spinosa.
Anagallis aquat. vulgaris folijs rotundioribus floribus cæruleis, sive Becabunga, Offic.
Gramen aquis innatans, Lob.
Ranunculus hederaceus, Lugd.

B 3 *Ra*

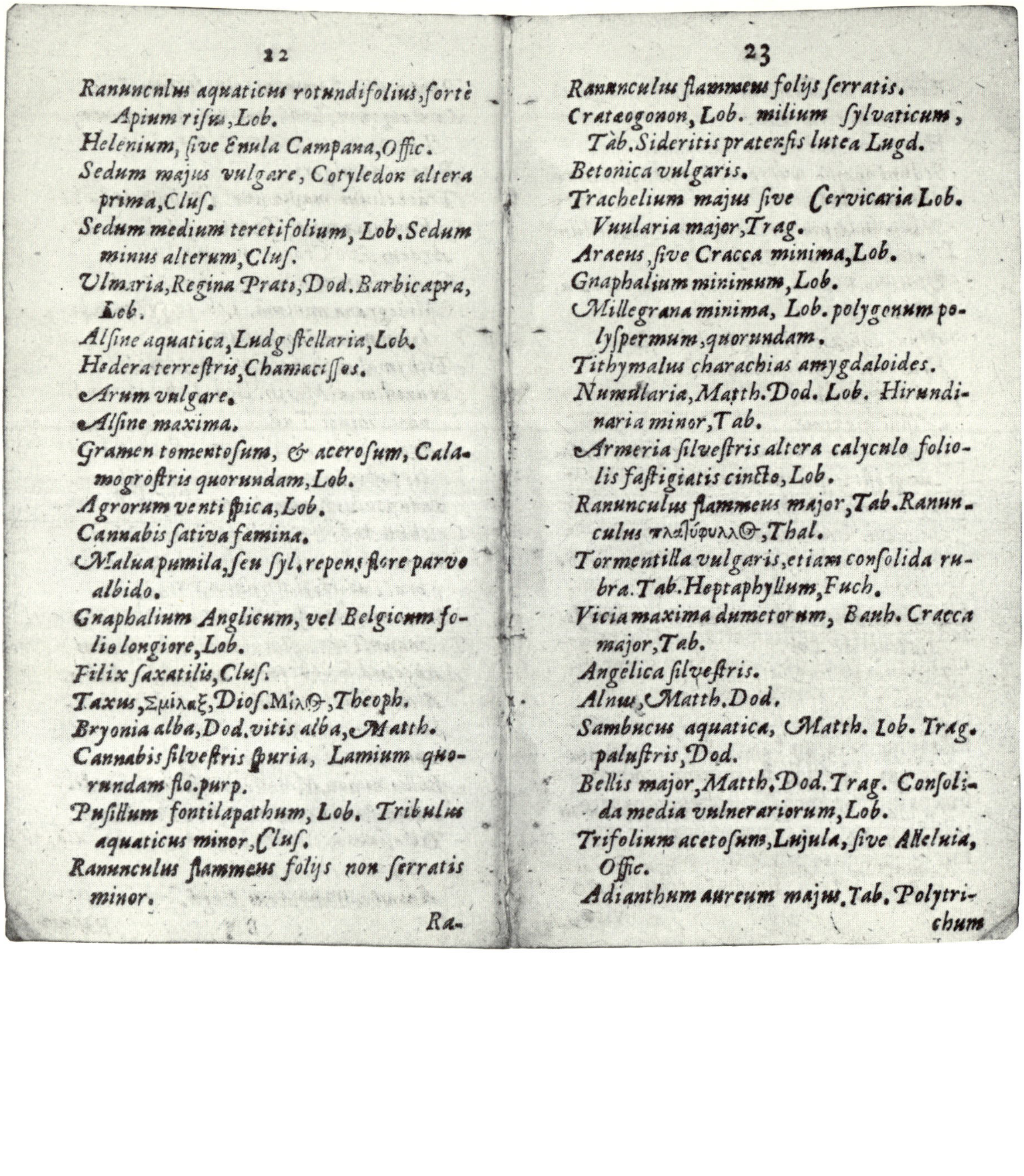

22

Ranunculus aquaticus rotundifolius, fortè Apium risus, Lob.
Helenium, sive Enula Campana, Offic.
Sedum majus vulgare, Cotyledon altera prima, Clus.
Sedum medium teretifolium, Lob. Sedum minus alterum, Clus.
Ulmaria, Regina Prati, Dod. Barbicapra, Lob.
Alsine aquatica, Ludg stellaria, Lob.
Hedera terrestris, Chamacissos.
Arum vulgare.
Alsine maxima.
Gramen tomentosum, & acerosum, Calamogrostris quorundam, Lob.
Agrorum venti spica, Lob.
Cannabis sativa famina.
Malua pumila, seu syl. repens flore parvo albido.
Gnaphalium Anglicum, vel Belgicum folio longiore, Lob.
Filix saxatilis, Clus.
Taxus, Σμίλαξ, Dios. Μίλος, Theoph.
Bryonia alba, Dod. vitis alba, Matth.
Cannabis silvestris spuria, Lamium quorundam flo. purp.
Pusillum fontilapathum, Lob. Tribulus aquaticus minor, Clus.
Ranunculus flammeus folijs non serratis minor.

Ra-

23

Ranunculus flammeus folijs serratis.
Crataeogonon, Lob. milium sylvaticum, Tàb. Sideritis pratensis lutea Lugd.
Betonica vulgaris.
Trachelium majus sive Cervicaria Lob. Vuularia major, Trag.
Araeus, sive Cracca minima, Lob.
Gnaphalium minimum, Lob.
Millegrana minima, Lob. polygonum polyspermum, quorundam.
Tithymalus charachias amygdaloides.
Numularia, Matth. Dod. Lob. Hirundinaria minor, Tab.
Armeria silvestris altera calyculo foliolis fastigiatis cincto, Lob.
Ranunculus flammeus major, Tab. Ranunculus πλατύφυλλος, Thal.
Tormentilla vulgaris, etiam consolida rubra. Tab. Heptaphyllum, Fuch.
Vicia maxima dumetorum, Bauh. Cracca major, Tab.
Angelica silvestris.
Alnus, Matth. Dod.
Sambucus aquatica, Matth. Lob. Trag. palustris, Dod.
Bellis major, Matth. Dod. Trag. Consolida media vulnerariorum, Lob.
Trifolium acetosum, Lujula, sive Alleluia, Offic.
Adianthum aureum majus. Tab. Polytrichum

chum majus, Trag. Fuch.
Adianthum aureum minus, polytrichum minus Trag. Fuch.
Musci capillaris altera species, Dod.
Leucojum Luteum, Dod. Keiri, Lob. & Offic. viola petraea lutea, Tab.
Flos solis Farnesianus, sive Aster Peruvianus, Col. } *ex horto.*
Sumach, sive Rhus Virginianum. }

Sole occidentem versus jam vergente, Cantuariam ingressi sumus nec diu morati, quin ad Cathedralem Archiepiscopatus Cantuariensis Ecclesiam nos contulimus, ibique sacris solito more peractis adfuimus; quibus finitis Templum quondam Thomæ Becketi delubro, cujus pars vilissima aurum fuit, famigeratissimum, circumspicimus; ibique sepulchra varia, sed præ alijs, magisque attente illud bellicosissimi Principis Edwardi cognomento Nigri, nec minore delectatione mirâ arte & elegantia picta fenestrarum vitra lustravimus. Tum turres conscendimus, & vrbem, ejusque in ruinam tendentes muros intuiti sumus. Nostrates enim veterum Spartanorum ritu plus præsidij in armis, quam muris ponunt.

Hic obvij facti sumus Carolo Anato viro,

virò, vt dixi, amicissimo, Richardique Iucksoni humanitate freti sumus, vt etiam societate, arte musica præstantissimi viri Guil. Pisinge, qui nobis non modo dum Cantuariæ fuimus, sed Londinum vsque socium se adjunxit.

Die dominicâ, quieti & divino cultui, vt consentaneum fuit, consecratâ, sequente aurora iterum assueto operi accingimur, plantæque sequentes inventæ sunt.

Teucrium pratense Lob. Verbenaca supina mas, Dod. Chamædrys silvestris, Clus. Ger.
Hieracium longius radicatum.
Hieracium montanum angustifolium primum, Tab.
Hieracium fruticosum latifolium glabrum, Bauh.
Veronica pratensis, Dod. veronica minor, Tab.
Hypericum pulchrum, Tragi.
Digitalis purpurea, Dod. Lob. Ephemeri Dios. alia species major, Col. Campanula syl. Trag.
Polygala flo. cæruleo, & flo. albo, Amarella Gesn. flos Ambervalis, Dod.
Ruscus, Bruscus, Oxymyrsine, myrtacantha, Lob.
Alsine aquatica folijs rotundioribus, sive Portulaca aquatica.

B 5 Argen-

Argentina, Dod. Lob. Potentilla, Matth. Anserina, Trag. Tab.

Serpillifolia aquatica, Bauh. sive Alsine palustris serpillifolia. Vide hujus fig. in fine lib. notatam, fig 5.

Saxifraga minor altera, flosculis albis, sem. nigro, forte, Alsine saxifraga angustifolia minima montana, Colum. Anglica saxifraga, Ad.

Cicutaria alba, Ludg. Cicutaria palustris flore candido, Cam.

Nullas præter has, & multas ante observatas, Cantuariam inter & Feversham vidimus. Illic morantes nostrûm aliqui mare versus tendentes eruunt.

Limonium vulgare sed perperam, vt etiam Behen rubrum.

Peucedanum vulgare seu Fœniculum porcinum.

Plantaginem aquaticam, Limonium verum Dios. vt alias ostendam.

Althæam vulgarem, sive Bismalvam, Offic. & Ger.

Tripolium vulgare, Amelli speciem palust. Col.

Allium sil vestre tenuifolium, Lob. Ophioscoridon, Matth.

Dein ad Officinam Nicholai Swayton nostræ artis socij tendimus, Ille nos benigne

nigne acceptos in hortum duxit, ibique crescebant herbæ sequentes.

Cucumer silvestris, Matth. Cucumer Asininus, Offic. Tab.

Aster atticus, sive Amellus, Matth. Tinctorius flos. 1. Trag. Aster Italorum purpurascens, Lob.

Leucoium purpureum,

Chondrilla rubra fœtida, Eyst.

Majorana.

Marum vulgare, Mastic, Gallorum & Anglorum. Ad. Tragoriganum Hispanicum primum, Clus.

Aquilegia.

Chamæcyparissus,

Asparagus vulgaris.

Carduus benedictus.

Cardiaca, Matth. Lob. Tab. melissa syl. Tragi.

Cameline, Myagrum alterum Thlaspi effigie, Lob.

Papaver flore albo multipl.

Cyclamen vernum, panis porcinus, Rapum seu Vmbilicus terræ, Quorund.

Parthenium, matricaria flore multiplice.

Caryophyllorum hortensium species variæ.

Polium luteum montanum, Tab. Ger.

Melilotus vulgaris, Offic.

Hisce visis, ille se nobis ducem prebuit, & nos ad sylvam varijs Plantis refertam ducebat, nce bat, vbi(omissis multis ex jam commemoratis) crescebant.

Blattaria flo. albis, Lob.

Androsæmum, Dod. Clymenon Italorum, sive Siciliana, Lob.

Holostium Ruellij, Gramen leucanthemū, Dod.

Cannabis fil. spuria, Lamium quorundam flo. albis.

Virga aurea, Matth. Dod. Symphytum petræum. 3. Tab.

Primula veris, sive Primula sylvarum, Lob. Alisma sylvarum, Col.

Oxalis, sive Acetosa vervicina tenuifolia, Lob. Oxalis ovina. Tab.

Gramen avenaceum variore grano nemorense, Lob.

Gramen nodosum avenacea pannicula.

Gramen Phalaroides majus, Lob.

Agrifolium Dod. Aquifolium, Matth. Lob.

Viola canina, cærulea inodora, sylvestris, serotina, Lob.

Gramen caninum medicatum vulgare.

Bugula, Dod. Lob. Consolida media, Matt. Fuch. Tab.

Orchis myodes prima.

Circæa Lutetiana, Lob. Lappa sylvestris, Trag Herba D. Stephani, Tab.

Scordium alterum Plinij, Lob. Salvia agre-stris,

stis, Dod. Sylvestris, Trag.

Polypodium.

Glaux vulgaris, Lob. Glycyrhiza sylvestris, Gesn. Fænugræcum syl. Trag.

Sylva relicta, gratijsq; amico actis, ingressi viam quæ Cantuaria Londinum ducit, qua potuimus celeritate per Sittingburne ad Rochester, sicq; tandem ad Gravesend venimus, & præter prius recensitas plantas, sese nostro conspectui hæ offerebant.

Muscus aquaticus denticulatus.

Alopecuros maxima Anglica paludosa, Lob.

Ruta muraria, Dod. Capillus veneris, Trag. Salvia vitæ, Lob. Ad. Lugd. Ad. Lugd. Adianthum album. Tab. paronychia, Matth.

Calamentha pulegij odore, Lob.

Quinquefolium petræum majus, Tab. Tormentillæ facie, Ger. Pentaphyllum album, Matth. exiguum alterum, Tragi.

Pilosella repens vulgaris.

Phyllitis Matth. Dod. Lingua cervina, Offic. Scolopendria vulgaris. Trag.

Aphaca, Lob. Dod. Elatine 3. Tab.

Antirrhinum medium.

Tanacetum vulgare, Matth. Dod. Artemisia Diosc. Tab.

Castanea

30

Castanea vulgaris.

Muscus terrestris pyxidatus Alabastriculos imitatus, Lob.

Synanchica, Lugd. Gallium montanum cruciatum, Col. Rubia genus parvum, Gesn. col.

Chamaecistus flore luteo. Helianthemum, Lob. Panax chironium, Matth. Hyssopus campestris, Trag.

Linum vulgare sativum.

Lathyrus latiore folio, Lob. Clymenum, Matth.

Saponaria Dod. Lob. Trag. Matth. Struthium, Fuch.

Coniza caerulea acris ; Tinctorius flos alter Tragi, Amellus mont. Col.

Tum Gravesendæ in opportunum tempus morati, primo æstu amnis ingressi splendidum, illic forte relictum actuarium naviogiolum octo remis actum, Londinum revertimur ; Deo opt. max. plurimas gratias agentes pro multis in nos collatis beneficijs ; eumque precamur, ut nostris & omnibus in publicum bonum susceptis laboribus optatum finem largiatur, *Amen*

FINIS.

31

Enumeratio Plantarum in Ericeto Hampstediano locisque vicinis crescentium.

In Ericeto Hampstediano, locisque ab euntibus & revertentibus visitatis crescentes plantas non in priore Itinerario recensitas hic enumerare operæpretium fore judico. Priore scripto nonnullas nobiscū vbique fere vulgares plantas denotavi, & alias non minus obvias hic nominare placet, namque ab hac arena inexercitatis & medicinæ rei, & remediorum id genus studiosis æque harum, quam aliarum cognitio, & nomina discenda sunt. Arbores aliquot ex industria omisi, vt qui nobiscum fructus gratia in viridarijs coluntur, & omnibus fere noti sunt, ut, Mali, Pyri, Cerasi, &c. Sic etiam frumenta, & legumina sativa, vt Siliginem, Triticum, Hordeum, Pisum, Fabam, &c. absque mentione præterij. Reliquos Arbores, etiam vbique obvios, consulto huc transtuli.

Synonima quædam plantis minus notis (si modo detur) aut varie ab Authoribus nuncupatis apposui. Sed ad rem ; magis obvias, & notas priore loco invenies.

Atri-

Atriplex silvestris latifolia, Lob. silvestris, 3. Math. Blitum 4. Trag. pes anserinus, Dod.

Atriplex olida, Lob. Canina, Tragi. Garosmum, Dod.

Alsine hederacea, Tab. Morsus gallinæ folio hederulæ, Lob.

Alsine minor, Tab. minima, Dod. Lob. minor multicaulis, Bauh.

Eupatorium cannabinum fæmina, Lob. Hepatorium aquatile, Dod. verbena supina, Trag.

Eruca sylvestris, Lob.

Sinapi silvestre minus bursæ pastoris folio, Lob. Sinapi. 3 Matth. an Irio Apulus alter levifolio Erucæ, Col.

Myrrhis sylvestris nova Æquicolorum, Col.

Paronychia vulgaris, Dod. alsinefolia, Lob. Bursa pastoria, 6. Trag.

Paronychia altera, Dod. rutaceo folio, Lob. Alsine petræa rubra, & Paronychia, 3. Tab.

Coniza minor, Trag. Matth. minima, Lob. Dod.

Smilax lævis, Matth. Dod. Volubilis major. Tab. Trag.

Lactuca syl. odore viroso folijs dissectis, Lactuca sylv. prior, Trag. silv. Matth.

Lactuca syl. alter odore magis viroso folijs

non

non dissectis, Lactuca agrestis odore opij Lob. Ad. Endivia major & 1. Trag. Thesion, Ludg.

Medica minor fructo cochleato aspero; Tribulus terrestris minor repens, Lugd. vt etiam ejus varietas folijs maculatis, Medica Arabica, Cam.

Rosa canina sylvestris inodora, Lob.

Rosa silvestris pomifera, Lob. Arvina, Tab.

Valeriana silvestris major.

Hyacinthus vulgaris, Anglicus & Belgicus, Lob.

Alliaria, Matth. Dod.

Potomogeiton majus vulgare, Matth. Trag.

Saxifraga Anglicana facie seseli pratensis, Lob.

Alectorolophos, pedicularis rubra, sive Ruta pedicularis, Tab. Fistularia. Dod.

Ranunculus hederaceus aquaticus, Lugd.

Gramen pratense vulgare.

Gramen loliaceum.

Gramen spicatum folijs vetonicæ caryophyllatæ, Lob.

Gramen palustre echinatum, Lob. aculeatum, Ludg.

Gramen tremulum, sive phalaris pratensis,

Gramen exile hirsutum cyperoides, Lob.

Gramen spartium capillaceo folio, minimum.

mum.

Gramen holostium minimum, fortè, Holostium Alpinum minimum, Bauh.

Gramen junceum sylvaticum, Tab. Ger.

Ranunculus bulbosus, Lob. tuberosus, Dod.

Chelidonium minus, sive scrophularia minor.

Fumaria vulgaris.

Fraxinus.

Quercus.

Rubus vulgaris.

Sambucus.

Betula, Matth. Lob.

Ligustrum.

Corylus sylvestris, Matth. nux avellana syl. Dod.

Erica vulgaris, Trag. Ger. prima, Matth. Dod.

Erica pumila calyculato vnedonis flore, Lob.

Muscus terrestris vulgaris.

Muscus in ericetis proveniens, Lob.

Muscus corallinus albicans, sive corallina montana, Tab.

Oxyacantha, Matth. Lob. spina appendix Plinij, Gesn.

Genistella, Trag. Dod. genistella aculeata, Lob.

Genista Tinctoria, Dod. Genistella infectoria, Lob.

Genista

Genista spinosa vulgaris.

Iuniperus vulgaris fruticosa.

Salix vulgaris longis & angustis folijs.

Salix latioribus albidis folijs.

Salix humilis repens, Lob. bombicifera.

Prunus sylvestris folijs latioribus, fructu majore.

Prunus sylvestris, Matth. Dod. Lob Acacia Germanica, Quorundam.

Populus Lybica, Matth. Dod. Cercis Theophrasti. Populus Lybica Plinij, Clus.

Carpinus, Matth. Ornus, Trag. Ostrys Theophrasti, Clus. Betulus, Lob.

Sorbus sylv. Alpina, Lob. Ornus, & fraxinus bubula, Dod. Fraxinea arbor, Trag.

Sorbus torminalis Plinij. Matth. Lob. Dod.

Alnus nigra baccifera, Ludg. Frangula, Matth.

Aria Theophrasti effigie Alni, Lob. Sorbus Aria cognominata, Clus.

Veronica vera & major, Lob. veronica mas, Dod. Fuch.

Cynosorchis morio, Lob. Orchis mas angustifolia, Fuch.

Orchis ornithophora, Lob.

Serapias candido flore montana maculatis folijs, Lob.

Palma Christi, Serapias fœmina pratensis, Ejusd.

Pri-

Primula pratenſis, Lob. Herba paralyſis vulgaris, Trag. Aliſma pratorum, Col.
Bulbocaſtanum, Dod. Nucula terreſtris ſeptentrionalium, Lob.
Chamæmelum, ſive Anthemis vulgatior, Lob.
Cardamine altera Lob. flos Cuculi, Dod. ſiſymbrium aquaticum alterum, Matth. Tab.
Cyperus gramineus, ſive miliaceus, Lob.
Cruciata, Dod. minor, Lob. Gallium. 2. Trag.
Alcea vulgaris, Dod Matth.
Lameum luteum, Lob. Tab.
Ervum ſylveſtre, Dod. ſilveſtre herbariorum, Lob. & Catanance, Quorundam.
Succiſa, Matth. Dod. Morſus diaboli, Trag. Lob.
Bifolium, Lob. Ophris, Matth.
Filix florida, vel. Oſmunda regalis, Lob. filix aquatica, Dod.
Lonchitis altera, Dioſcoridis, Lob. Aſplenium ſilveſtre, Trag. Lonchitis aſpera minor, Matth.
Vaccinia nigra, Dod. Myrtillus, Matth. Vitis Idæa vulgaris, Cluſ. Idæa. 1. Tab.
Iuncus bombycinus, Lob. Linum pratenſe, Geſ. Gnaphalium, Trag. Gramen tomentarium, Ger.
Trifolium paludoſum, Lob. Ger. Limonium,

um, Cordi. menianthes paluſtre, Ludg.
Ros ſolis, Dod, Rorella prima, Tab.
Muſcus terreſtris, Matth. Dod. Muſcus clavatus, Lob. Ger. Lycopodium, Tab.
Orobanche, Rapum geniſta.
Lilium convallium, Trag. Matth.
Scabioſa montana minima, Lob.
Lyſimachia galericulata, Lob. ſideritis cærulea, Thal, Tertianaria. Tab. Gratiola, latifolia, & Cærulea, Quorundam.
Lyſimachia galericulata minor, ſive Graticla latifolia, Ger.
Ornithopodium perpuſillum, Lob.
Anagallis flore luteo, Ejuſd.
Serratula, Matth. Dod. Centauroides, vel centaurium majus ſylveſtre Germanicum, Thal.
Aſtragalus ſylvaticus. Thal.
Cirſium Anglicum, Lob.

FINIS.

38

Iconum Declaratio.

1. 2. FVcus longiss. latissimo, crassoq; folio. Bauh. &c mire variat nonnunquam enim profert ex radice fibrosa, & pediculo semipedale unicum folium tantæ longitudinis, ut ulnam æquet, & tum fere est palmaris latitudinis, ita ut belle ancipitem gladium representet. Aliquando duo folia prius descripto, excepta magnitudine, similia promit. Nonnunquam Polyschides sive multifidum folium emitit quod in 8. 9. 10. aut 12. plures, aut pauciores partes dividitur. Aliquando etiam unum integrum, & alterum multifidum folium ex eadem radice prodeunt, ut præmissâ figurâ ostenditur. Ejus ideo pars fig. 1. notata sit Fucus Phasganoides; & altera, fig. 2. signata, Polyschides.

Hactenus de formâ, nunc de colore, qui omnibus idem non est, namq; magis virescunt alij, & hi vix siccari possunt; alij albicant & statim siccantur, & tum pergamenæ adeo sunt colore & consistentia similes, ut pro ea, ab ijs qui hoc ignorant, sumi possint. Rite describitur dum adhuc crescit ejus prima varietas, ut & reliquæ a Bauhino, in Prod. lib. 10, cap. 8. num. 1. 2. sed præcipue 3. sub nomine Fuci Longiss. &c.

3 F[illegible]cus

39

3. Fucus spongiosus nodosus, *Ang. Sea Ragged-staffe.*

Hæc valde succulenta, & fungosa planta est, pollicaris crassitudinis; flavescit obscure & multis inæqualibus protuberat appendicibus, aut nodis, unde rite a Guardiano nostro, *Tho. Hickes*, vocata fuit Anglicè, *Sea Ragged-staffe.* Crescentem non vidimus, sed pedalis longitudinis, unam & alteram plantam invenimus.

4. Alsine aquatica folijs rotundioribus, sive Portulaca Aquatica.

Describitur hæc a *Casp. Bauhino* in Prodromo, pag. 118. sub nomine, Alsines palustris minoris folio oblongo.

5. Serpillifolia aquatica. *Io. Bauhini*, &c.

Describitur cum priore a *Casp. Bauhino* loco citato, sub nomine, Alsines palustris minoris serpillifoliæ. Depingitur etiam a *Io. Bauhino* (sed non describitur) in Catalogo plantarum circa Balneum Bollense. pag. 194.

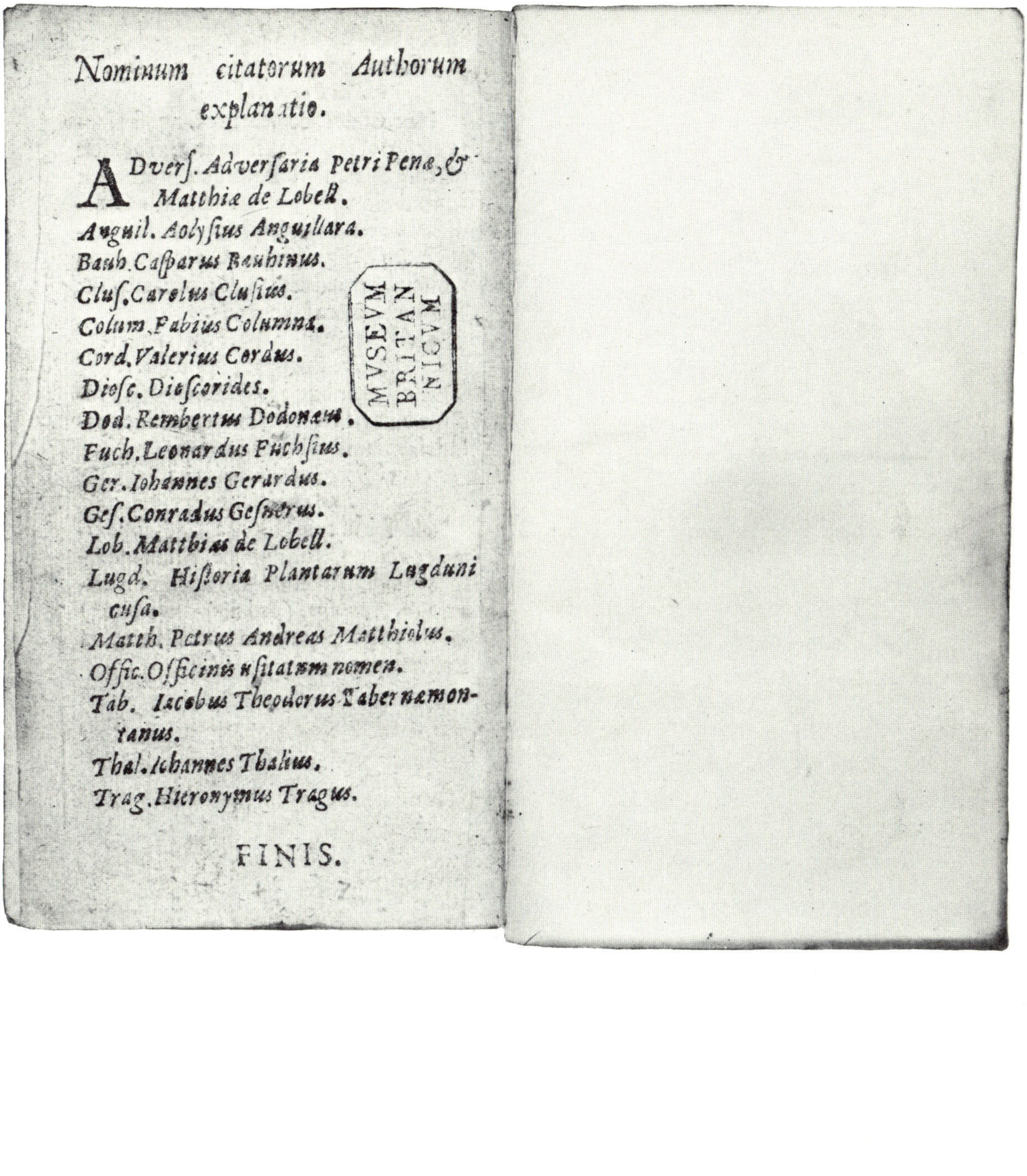

Nominum citatorum Authorum explanatio.

ADvers. Adversaria Petri Penæ, & Matthiæ de Lobell.
Anguil. Aolysius Anguillara.
Bauh. Casparus Bauhinus.
Clus. Carolus Clusius.
Colum. Fabius Columna.
Cord. Valerius Cordus.
Diosc. Dioscorides.
Dod. Rembertus Dodonæus.
Fuch. Leonardus Fuchsius.
Ger. Iohannes Gerardus.
Ges. Conradus Gesnerus.
Lob. Matthias de Lobell.
Lugd. Historia Plantarum Lugduni cusa.
Matth. Petrus Andreas Matthiolus.
Offic. Officinis usitatum nomen.
Tab. Iacobus Theodorus Tabernæmontanus.
Thal. Iohannes Thalius.
Trag. Hieronymus Tragus.

FINIS.

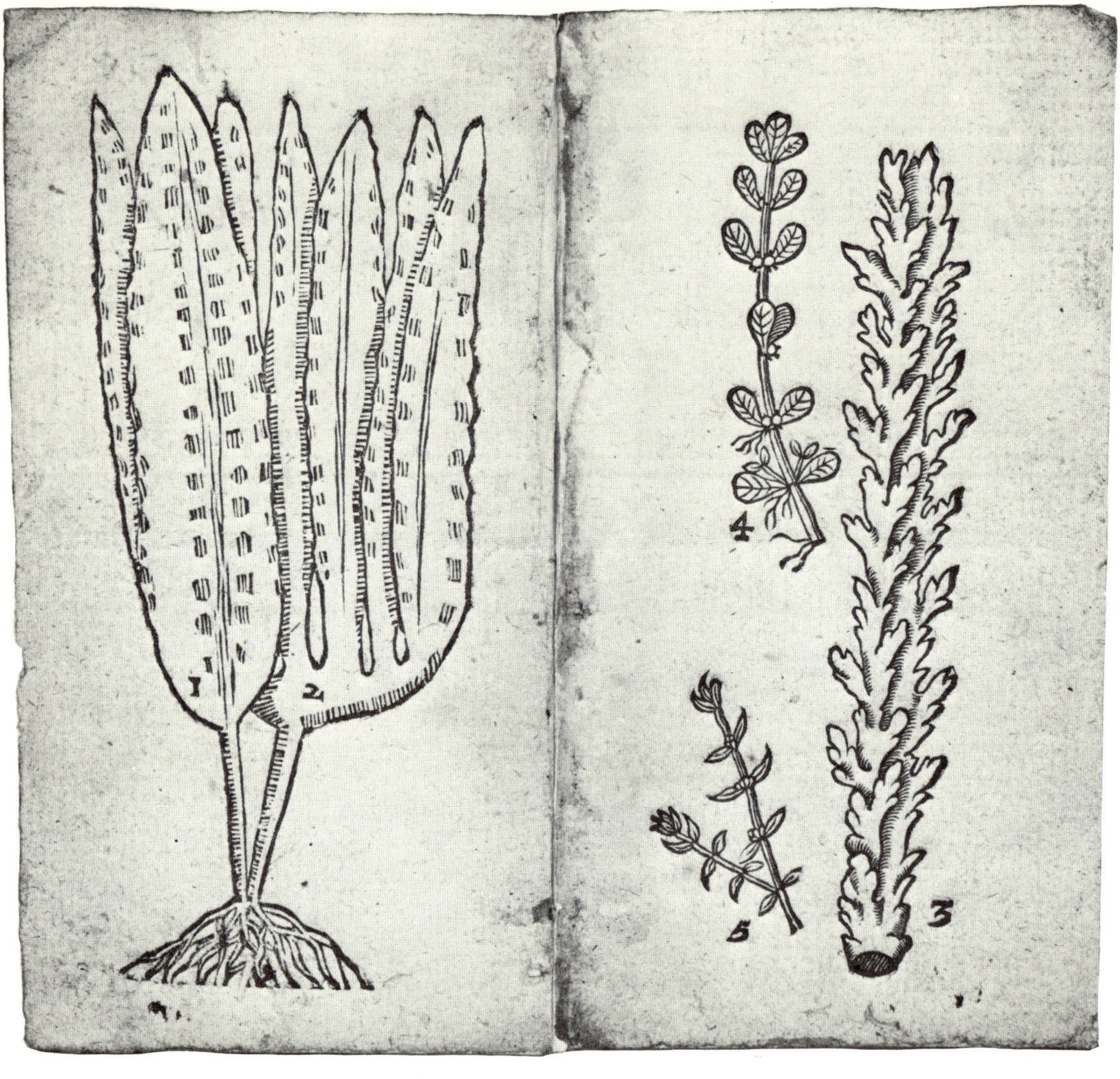
1
2
4
5
3

Translation of Thomas Johnson's Descriptio Itineris, 1632

By
CHARLES E. RAVEN
assisted by
RONALD E. LATHAM
Modern Scientific Names by
FRANCIS ROSE
assisted by
MAURICE BURTON (*"Sea Serpent"*)
J. S. L. GILMOUR
W. MARGADANT (*Bryophytes and Lichens*)
J. H. PRICE (*Marine Algae*)
P. W. RICHARDS (*Bryophytes and Lichens*)
and
H. G. VEVERS (*Animals*)

Each entry in the plant lists consists of Johnson's name, followed, in square brackets, by the modern scientific name, or a discussion of this when there is doubt. Some obvious mistakes in Johnson's names have been corrected, and some of his spelling modernised.

At the foot of each page of the translation, after the page number of the continuous pagination, the relevant page number or numbers of the facsimile are given thus: Facs. p. oo.

Footnotes signed R. are by Charles E. Raven, those signed L. by Ronald E. Latham, F.R. by Francis Rose, and G. by J. S. L. Gilmour.

In modern editions of the *Bacchides*, the passage on the title page (III, 6, line 15) appears with a slightly different reading (*ignavia* for *ignavi*); the translation is intended to convey the sense in which Johnson appears to have understood it (i.e. with the implication that even such an insignificant person as himself had need to guard against detraction). L.

For a discussion of the printer's device, see Introduction, p. 5.

The Description of a Journey
Undertaken for the Discovery of Plants
Into the County of Kent
In the Year of Our Lord 1632

And

A List of Plants growing on
Hampstead Heath
And its Neighbourhood

Even the lowest of the low
Do well to parry envy's blow.
[*Plautus, Bacchides*

PRINTED BY THOMAS COTES, 1632
(Facs. p. 71)

TO MEN HIGHLY REGARDED AND WIDELY CONSULTED,
BY THEIR ART AND EXPERIENCE MEN OF RENOWN AND DIGNITY,
THE RECTOR OR MASTER, WARDENS AND THEIR ASSISTANTS
OF THE RENOWNED SOCIETY OF LONDON APOTHECARIES,
THIS THEIR WORK, FOR WHAT IT IS WORTH, AS
A TOKEN OF ESTEEM AND AFFECTION
IS PRESENTED AND DEDICATED BY
THE TRAVELLING FELLOWS.

(Facs. p. 72)

To Friendly Readers

TWO YEARS AGO[1] a journey was undertaken according to our custom for the purpose of discovering plants; it was described with a hurrying pen; for the sake of my friends I sent this description to the press. It was amazing, although the work was privately printed for the sake of a few readers, what a variety of praise and blame it received. For some folk not only ridiculed our labour as vain and superfluous but derided all more precise knowledge of plants as useless, supposing that it is enough to know them merely by name and from reading. It is very certain that the men of old who founded medicine were not so indolent and ignorant. This is easy to prove, even if we omit the testimony of Galen and others both ancient and modern, by quoting Oribasius. His words are these: 'The knowledge of the "simple" drugs and their virtues is so absolutely necessary that no one without it can practise medicine correctly' [this is quoted in Greek and then rendered in Latin]. He said that knowledge of 'simples' was not merely necessary but absolutely necessary. But so far as this affects the careless druggists who so stupidly ridicule us, I will expose and depict in a few words their supreme ignorance, which springs from carelessness or arrogance. Almost every day in the herb market one or other of them, to the great peril of their patients, lays himself open to the mockery of the women who deal in roots. These women know only too well the unskilled, and thrust upon them brazenly what they please for what you will. Not a few of them buy from the women Cotyledon aquatica acris [*Hydrocotyle vulgaris* L.] for Umbilicus veneris [*Umbilicus rupestris* (Salisb.) Dandy]; Sedum minimum Diosc. 3 [*Sedum acre* L.] for Sedum minus (i.) vermiculare, Offic. [probably *Sedum anglicum* Huds.]; Polygala flore cæruleo [*Polygala vulgaris* L.] for Gratiola [*Gratiola officinalis* L.]; Anagallis flo. phænicio [*Anagallis arvensis* L.] for pimpinella sanguisorba [*Sanguisorba officinalis* L.] or pimpinella saxifraga [*Pimpinella major* (L.) Huds.]: and with less risk but no less palpable error Sambucus [*Sambucus nigra* L.] for Ebulus [*Sambucus ebulus* L.]; Bonus Henricus [*Chenopodium bonus-henricus* L.] for Mercurialis [*Mercurialis perennis* L.]; filix mas [*Dryopteris filix-mas* agg.] for filix florida (i.) osmunda regalis [*Osmunda regalis* L.]. I could name not a few others; but these are more than enough to convict these fellows as guilty of gross ignorance. Is not the fate

[1] Johnson is clearly referring to the journey in 1629 described in his *Iter*. 'Two years ago' must therefore be a slip, or possibly he deliberately did not count 1632, the year in which he was writing. G.

(Facs. pp. 73–4)

of patients who rely upon the help of such doctors and druggists pitiable? For the doctor relies on the druggist and the druggist on a greedy and dirty old woman with the audacity and the capacity to impose anything on him! So it often happens that the patients' safety depends on the herbal knowledge of an ignorant and crafty woman. So for the future let them cease to deride and denigrate as vain and useless the labours of others undertaken for the public good, when they are themselves deserving not only of censure but of heavier punishment.

These are rejoinders that I had to make in our defence, so that the pleasure they took in speaking evil they may lose by hearing evil. But after this, in place of these calumnies, we hope to arouse only such strife as that about which Hesiod writes this commendation, 'This strife is good for mortals'.[1] This for the sake of yourselves and of the Commonwealth is the heartfelt wish of Thomas Johnson.

[1] Hesiod, *Works and Days*, l, 24. L.

(Facs. p. 74)

Description of a Journey undertaken for the Discovery of Plants into the County of Kent, 1632

THIS YEAR when the Society of London Apothecaries was discussing its annual pilgrimage for the sake of finding plants, the Warden of our Society, Thomas Hicks,[1] knowing that we had for several years past been in the habit of undertaking a journey of three or four days for the sake of traversing the natural habitats of the plants, not only encouraged us as usual to spend some days on it, but promised himself to take his fair share of the work and more than his share of the expenses. He found no difficulty in persuading a willing audience. We agreed forthwith, and thanked him for thinking us worthy of such an honour.

So on the first of August those who were ready for their part, William Broad, Leonard Buckner, Robert Larkin, James Clarke,[2] and Thomas Johnson, assembled and entered the house of Thomas Hicks our Warden. Hence, when we had first taken breakfast by his munificence and no small provision for the way, we went to the barge, boarded it, spread our sails to the wind, and left London.

In the course of ten hours we covered sixty English miles, so that at the first approach of night we put to shore by the chalk cliffs of the isle of Thanet and the bay of Margate, where piles have been driven in and tied together and rocks thrown into the water so as to make a mole or pier for the more convenient berthing of ships.[3] Here we left our own ship and went to our inn, where we found everything fit and ready, including our most attentive host, whose name is Richard Pollard. While we were in the island, he never left us, nor did he, like most others, plunder us when we departed. Here after a little refreshment we took ourselves off to bed.

Next morning we sallied forth as far as a fort set on a steep promontory, fortified by nature more than by art, and collected on the shore and on the steep cliffs the plants whose names follow:

[1] For Thomas Hicks, Upper Warden of the Society, see Underwood (1963). L.

[2] James Clarke, who died in 1652 when holding the office of Renter Warden of the Society, was among those mentioned in the preface to Johnson's edition of Gerard's *Herball*, as also was Hicks. R. & L.

[3] The pier was ancient and ill-repaired (Hasted, **10**, 315–19). R.

(Facs. pp. 75–6)

Bryon Lactucæ foliis, Lob. Lactuca marina, Taber. [*Ulva lactuca* L.] See Appendix 1, p. 139.

Alga membranacea ceranoides, Bauh. [*Rhodymenia palmata* (L.) Grev.] See Appendix 1, p. 139.

Fucus longissimo, latissimo, crassoque folio, Bauh. prodr. forte Phasganoides ejusdem, nomen enim rei bene convenit, ut etiam, Alga latifolia coriacea, ejusdem, & fere affirmare ausim, si non eædem, ejusque speciei varietates sint. Vide figuram hujus in fine lib. notatam, fig. 1. & varietat. polyschid, fig. 2. [*Laminaria saccharina* (L.) Lamour, and *L. digitata* (Huds.) Lamour] See Appendix 1, p. 140.

Fuci ejusdem varietas polyschides. [*Laminaria digitata* (Huds.) Lamour] See Appendix 1, p. 140.

Alga membranacea ceranoides, Bauh. [*Rhodymenia palmata* (L.) Grev.] See Appendix 1, p. 139.

Fucus marinus quartus. Dod. [*Halidrys siliquosa* (L.) Lyngb.] See Appendix 1, p. 140.

Fucus spongiosus ramosus. [Possibly *Dumontia incrassata* (O. F. Müll.) Lamour] See Appendix 1, p. 141.

Fucus spongiosus nodosus. Cujus figura notatur, fig. 3. [Possibly *Dumontia incrassata* (O. F. Müll.) Lamour] See Appendix 1, p. 141.

Fucus ferulaceus, Lobelii. [Probably *Cystoseira foeniculacea* (L.) Grev., but possibly *Cystoclonium purpureum* (Huds.) Batt.] See Appendix 1, p. 142.

Quercus marina vulgaris foliis non serratis. [*Fucus vesiculosus* L.] See Appendix 1, p. 142.

Quercus marina foliis serratis. [*Fucus serratus* L.] See Appendix 1, p. 142.

Corallina ramosa parva, & pennata longior. [Probably *Corallina* spp.] See Appendix 1, p. 143.

Gramen caninum geniculatum marinum, Lob. [*Agropyron junceiforme* (A. & D. Löve) A. & D. Löve]

Gramen parvum marinum spica loliacea. [*Agropyron pungens* (Pers.) Roem. & Schult., or possibly the hybrid with *A. junceiforme* (*A.* × *acutum* auct.)]

Cornu cervinum hirsutum, Coronopus, sive Herba stella. Mat. Lob. [*Plantago coronopus* L.]

Tragon improbus, Matth. [*Salsola kali* L.]

Papaver corniculatum flore luteo. [*Glaucium flavum* Crantz]

Papaver erraticum minus. [Probably *Papaver dubium* L.]

Bellis minor vulgaris. [*Bellis perennis* L.]

Alsine corniculata, Clusii. [*Agrostemma githago* L., see Johnson's Gerard, p. 616]

(Facs. p. 76)

Anthyllis leguminosa Belgarum, Lob. Lagopodium. Tab. [*Anthyllis vulneraria* L.]

Trifolium fragiferum, Clus. floribus dilute rubentibus. [*Trifolium fragiferum* L.]

Trifolium luteum minimum, Lob. sive medica sem. racemoso. [Johnson's nomenclature of the small yellow *Trifolium* and *Medicago* species is difficult to interpret. Fl.K. gives *T. dubium* Sibth. for this entry, and *M. lupulina* L. for the next, but *M. lupulina* and *T. campestre* Schreb. respectively seem more likely]

Trifolium luteum majus lupulinum, sive lupulus silvaticus, Thal. [Probably *Trifolium campestre* Schreb.; see previous entry]

Trifolium corniculatum, Dod. Melilotus coronata, Lob. minor glabra. [*Lotus corniculatus* L.]

Pimpinella Saxifraga major, Dod. Lob. Tragoselinum majus, Tab. [*Pimpinella major* (L.) Huds.]

Pimpinella saxifraga minor, sive Bipinilla Lob. [*Pimpinella saxifraga* L.]

Onopyxos Dodonæi, Lugd. [Fl.K. gives *Carduus nutans* L., but Johnson (*Mercurius*) uses another name for this, and for *Onopordon*. The figure in Gerard is difficult to interpret, and the identification of this entry remains doubtful]

Carduus stellaris, Dod. seu Calcitrapa, Ejusd. & Lob. [*Centaurea calcitrapa* L.]

Rheseda vulgaris, eruca perigrina, Italica vel cantabrica, Lob. [*Reseda lutea* L.]

Sinapi sativum vulgare. [*Brassica nigra* (L.) Koch]

Sinapi alterum siliqua falcata, sive sinapi alterum sativum, Lob. [Probably *Sinapis alba* L.]

Vrtica non mordax vulgaris purpurea. [*Lamium purpureum* L.]

Vrtica silvestris asperior, sive Urtica fœmina. [*Urtica dioica* L.]

Vrtica minor acrior. [*Urtica urens* L.]

Cotula fœtida vulgaris. [*Anthemis cotula* L.]

Cotula fœtida flore elegante multiplice. [Possibly *Matricaria recutita* L. (*M. chamomilla* auct.)]

Gallium luteum. [*Galium verum* L.]

Spumeum papaver, Lob. vulgo behen album; Herba articularis, Tab. [*Silene vulgaris* (Moench) Garcke (*S. cucubalus* Wibel)]

Horminum sylvestre, sive Oculus Christi, est Hormini silvestris, 4. species 5. Clus. Hist. [*Salvia horminoides* Pourr. (*S. verbenaca* auct. Brit.)]

Bursa pastoris major & minor. [*Capsella bursa-pastoris* (L.) Medic.]

Dens leonis, caput Monachi, rostrum porcinum, Taraxacon major & minor. [*Taraxacum officinale* Weber]

(Facs. p. 77)

Trifolium pratense album. [*Trifolium repens* L.]
Scabiosa minor, sive columbaria. [*Scabiosa columbaria* L.]
Gramen Typhinum minus. [*Phleum bertolonii* DC. (*P. nodosum* auct.)]
Caucalis nodoso echinato semine, Bauh. [*Torilis nodosa* (L.) Gaertn.]
Rubia sylvestris. [*Galium mollugo* L.]
Acanthium sylvestre, sive Onopordon. [*Onopordon acanthium* L.]
Lycopsis Anglica, Lob. Echii altera spec. Dod. [*Echium vulgare* L.]
Atriplex marina repens, Lob. [Probably *Atriplex laciniata* L., or *A. glabriuscula* Edmonst.]
Atriplex silvestris polygoni, sive helxines foliis. [*Atriplex patula* L.]
Helxine, sive parietaria, abundant in hedges. [*Parietaria diffusa* Mert. & Koch]
Elaphoboscum erraticum, Tab. sive pastinaca sylvestris, Dod. [*Pastinaca sativa* L.]
Irio, sive Erysimum vulgare Lob. Verbena fœm., Trag. [*Sisymbrium officinale* (L.) Scop.]
Anonis, sive ononis non spinosa. [*Ononis repens* L.]
Lapathum acutum vulgare. [*Rumex* sp. The pre-Linnean names and figures are difficult to interpret and the identity of this entry is doubtful]
Alsine media glabra. [*Stellaria media* (L.) Vill.]
Artemisia vulgaris. [*Artemisia vulgaris* L.]
Euphrasia rubra sive secunda, Dod. Odontitis, Tab. Sideritis pratensis rubra. Lugd: Cratægonon Euphrosine, Ger: [*Odontites verna* (Bell.) Dum.]
Luteola, Lob. Pseudostruthium, Matt. Antirrhinon, Trag. catanance, Tab. [*Reseda luteola* L.]
Daucus silvestris vulgaris. [*Daucus carota* L.]
Rapistrum arvorum, Lob. [*Sinapis arvensis* L.]
Cicuta vulgaris. [*Conium maculatum* L.]
Cicutaria fatua, Lob. Petroselini vitium, Tragi. [*Aethusa cynapium* L.]
Ballote, sive Marrubium nigrum. [*Ballota nigra* L.]
Sedum minimum, sive 3 Dios. Illecebra, Dod. [*Sedum acre* L.]
Erigeron, sive senecio minor vulgaris. [*Senecio vulgaris* L.]
Erigeron tomentosum alterum, Lob. [Probably *Senecio sylvaticus* L., rather than *S. viscosus* L.]
Hieracium montanum saxatile, Columnae. [*Leontodon taraxacoides* (Vill.) Mérat (*L. leysseri* Beck.)]

Then returning to the inn we waited for Simon Rose, who practises medicine in the place, so that we might have him as a guide to the island in our

(Facs. pp. 77–8)

search for plants. He came after a short delay and led us, but not by the public path, to Nash. On the way the following plants were collected:

Telephium vulgare, Fabaria Matth. Faba inversa, Lob. [*Sedum telephium* L.]
Hordeum spontaneum spurium, Holcus Plinii, Anguil. [*Hordeum murinum* L.]
Iacobæa vulgaris major. Senecio major. Matth. [*Senecio jacobaea* L.]
Iacobæa minor foliis magis dissectis. [*Senecio erucifolius* L.]
Gnaphalium vulgare, filago, centunculus. [*Gnaphalium uliginosum* L.]
Anagallis flore phoeniceo. [*Anagallis arvensis* L.]
Ranunculus vulgaris pratensis repens. [*Ranunculus repens* L.]
Ranunculus vulgaris pratensis surrectis cauliculis. [*Ranunculus acris* L.]
Sophia Chirurgorum, Lob. Seriphium Germanicum, Tragi, Thalictrum, Tab. [*Descurania sophia* (L.) Webb ex Prantl]
Gramen cristatum, Bauh. [*Cynosurus cristatus* L.]
Lapathum acutum minimum, Lob. [*Rumex conglomeratus* Murr.]
Trifolium pratense purpureum. [*Trifolium pratense* L.]
Tithymalus helioscopius. [*Euphorbia helioscopia* L.]
Cyanus minor vulgaris, Baptisecula, Trag. [*Centaurea cyanus* L.]
Iacea nigra vulgaris. [*Centaurea nigra* L., probably subsp. *nemoralis* (Jord.) Gugl.]
Iacea major, Lob. flos Cyanoides, Dod. [*Centaurea scabiosa* L.]
Helxine cissampelos altera atriplicis effigie, Lob. volubilis nigra, Tab. Orobanche, Tragi. [*Polygonum convolvulus* L.]
Chrysanthemum segetum, Lob. Clus. Bellis major lutea, Trag. [*Chrysanthemum segetum* L.]
Elatine Dioscor. sive veronica fœmina, Fuch. & Dod. [*Kickxia spuria* (L.) Dum.]
Elatine, Matth. Elatine altera, Dod. Lob. [*Kickxia elatine* (L.) Dum.]
Verbena. [*Verbena officinalis* L.]
Cuscuta, Matth. Androsace, Tragi. [*Cuscuta epithymum* (L.) L., or *C. europaea* L.]
Mentha Cattaria, Lob. Nepeta vulgaris, Trag. [*Nepeta cataria* L.]
Asperula cærulea arvensis. [*Sherardia arvensis* L.]
Myosotis scorpioides, Lob. Euphrasia cærulea, Tragi, & Tab. [The names suggest *Myosotis scorpioides* L.; but the dry habitat makes *M. arvensis* (L.) Hill more likely]
Ladanum segetum Plinii, & Tetrahit angustifolium aliis, Ludg. Alyssum Galeni flore purpureo, Tab. [*Galeopsis angustifolia* Hoffm.]
Coniza media Matth. Calaminthæ 3. genus, Fuch. [*Pulicaria dysenterica* (L.) Bernh.]

(Facs. pp. 78–9)

Coniza major Matth. Baccharis monspeliensium, Lob. [*Inula conyza* DC.]

Millefolium vulgare floribus albis, & flo. Carneis. [*Achillea millefolium* L.]

Scabiosa major satorum. [*Knautia arvensis* (L.) Coult.]

Scabiosa floribus albis. [*Knautia arvensis* (L.) Coult., white form.]

Pseudomelanthium, nigellastrum, lychnis segetum. [*Agrostemma githago* L.]

Marrubium aquaticum acutum, Gerardi, & panax Coloni ejusdem, Stachys palustris, Ges. Sideritis gravis odoris, Thal. [*Stachys palustris* L.]

Calamintha vulgaris, Offic. [*Calamintha ascendens* Jord.]

Lampsana, Lob. Sonchus sylvaticus, Tab. & Ger. [*Lapsana communis* L.]

Hieracium minus præmorsa radice, Lob. Hieracium leporinum, Ger. [*Leontodon autumnalis* L.]

Hieracium intubaceum, Tab. [Hieracium sp. probably *H. umbellatum* L.]

Lichen, hepatica vulgaris. [Possibly a thallose liverwort (e.g. *Pellia* sp.), though a lichen such as *Peltigera* sp. is more likely in dry Thanet. *Marchantia polymorpha* is another possibility]

Virga pastoris, Matth. Lob. Dipsacus tertius Dod. [*Dipsacus pilosus* L.]

Caryophyllata vulgaris. [*Geum urbanum* L.]

Plantago major vulgaris. [*Plantago major* L.]

Plantago minor quinquenervia, Lob. [*Plantago lanceolata* L.]

Plantago pannicula sparsa, sive scoparia, Lob. ['Besome Plantaine, or Plantaine with spoakie tufts' of the *Mercurius*; presumably *Plantago lanceolata* L. var. *timbalii* Reich. f., which is characteristic of cultivated land]

Serpillum. [*Thymus* sp., probably *T. pulegioides* L., which is commoner than *T. drucei* Ronn. in Kent]

Esula exigua,Trag. Tithymalus leptophyllos, Matth. [*Euphorbia exigua* L.]

Alsine sive Hippia major, Lob. [Probably *Cerastium arvense* L.]

Alsine foliis Trissaginis, Lob. Morsi Gallinæ, 3 Trag. [Probably *Veronica polita* Fr., which is more likely on chalk than *V. agrestis* L.]

Alsine foliis veronicæ, Tab. Elatine polyschides, Lugd. [*Veronica arvensis* L.]

Peirchpeir Anglorum. Lob. Alchimilla montana, Col. [*Aphanes arvensis* agg., probably *A. arvensis* L., *sensu stricto* on chalk]

Morsus gallinæ folio hederulæ alter, Lob. Ballote crispa minor, Lugd. [*Lamium amplexicaule* L.]

Vrtica heraclia, Tragi. Herculea, Tab. Galiopsis Dioscoridis, Clus. [*Galeopsis tetrahit* L.]

Lychnis silvestris flo. albo. [*Silene alba* (Mill.) E. H. L. Krause]

[*Note.* The plants so far included in this list tend to check one another as to

(Facs. p. 79)

identification, as they present a consistent picture of the flora of chalky cornfields and roadside banks such as one finds today in those parts of Thanet not urbanized. F.R.]

Eupatoreum vulgare, sive Agrimonia, Offic. [*Agrimonia eupatoria* L.]
Caucalis semine aspero flosculis subrubentibus, Bauh. [Fl.K. suggests *Torilis japonica* (Houtt.) DC. but, if in corn, *T. arvensis* (Huds.) Link seems equally possible]
Myrrhis silvestris, sive Cerefolium silvestre, Tab. [*Anthriscus sylvestris* (L.) Hoffm. seems intended, but in August *Chaerophyllum temulentum* L. is more likely]

[*Note*. We seem now to be passing along shady hedges, away from the open country to the North. F.R.]

Prunella vulgaris, Tragi. Consolida minor, Matth. [*Prunella vulgaris* L.]
Periclymenum non perfoliatum, Caprifolium, Offic. [*Lonicera periclymenum* L.]
Viburnum, Matth. viurna vulgi gallorum & Ruellii, Lob. [*Viburnum lantana* L.]
Geranium Rupertianum, sive herba Roberti. [*Geranium robertianum* L.]
Geranium alterum Dioscor. sive Columbinum. Tab. pes columbinus, Dod, & ejus altera species foliis majoribus & magis dissectis. [This entry presumably includes two species. The first is probably *Geranium molle* L. or *G. pusillum* L., and the second either *G. dissectum* L. or *G. columbinum* L., which Johnson does not seem to have distinguished.]
Evonymus Theophrasti, Lob. Dod. Carpinus Theophrasti, Trag. [*Euonymus europaeus* L.]
Hedera vulgaris. [*Hedera helix* L.]

[*Note*. Here we enter more open country again. F.R.]

Pecten Veneris, Matth. Scandix, Dod. [*Scandix pecten-veneris* L.]
Clinopodium vulgare, Matth. Origanum, 4. Tragi. [*Clinopodium vulgare* L.]
Sonchus arborescens, Tab. Ger. [*Sonchus arvensis* L.]
Gramen spica Brizæ majus, Bauh. [Probably *Brachypodium sylvaticum* (Huds.) Beauv., but possibly *B. pinnatum* (L.) Beauv.]

So at last we came to Nash and were kindly welcomed by the squire of the place.[1] In his garden were growing:

Draco hortensis, sive Draco herba, Dod. [*Artemisia dracunculus* L.]

[1] Nash Court, some three miles south of Margate, was then in the possession of William Cleybrooke, Esquire, whose father Paul had bought it in 1622 (Hasted, **10**, 341). R.

(Facs. pp. 79–80)

Raphanus rusticanus. [*Armoracia rusticana* Gaertn., Mey. & Scherb.]

Antirrhinum majus purpureum. [*Antirrhinum majus* L.]

Cataputia minor, Lob. Lathyris, Matth. Trag. Dod. [*Euphorbia* sp., probably *E. lathyrus* L.]

Sisarum, sive siser, Matth. [*Sium sisarum* L.]

Levisticum vulgare, Dod. Ligusticum, Tab. [*Levisticum officinale* Koch.]

Malua hortensis rosea. [*Althaea rosea* Cav.]

Hepatica nobilis, Trag. Herba Trinitatis, Matth. [*Anemone hepatica* L.]

Rha rotundifolium, sive Hippolapathum, rotundifolium, Lob. [*Rheum rhaponticum* L.]

Sedum serratum flore rubente maculato, Parkinsoni. [*Sedum* sp.]

Auricula vrsi, Matth. Lunaria arthritica, Gesn. Alysma, seu Damasonium, Dioscoridis, Col. [*Primula* sp., probably *P. auricula* L.]

Helleborus niger verus flo albo. [*Helleborus niger* L.]

Flos solis major, sive Chrysanthemum Peruvianum. [*Helianthus annuus* L.]

Melissa vulgaris. [*Melissa officinalis* L.]

Napellus verus cæruleus, Ad. Lob. Aconitum cæruleum, Generi. [*Aconitum napellus* agg.]

Asarum. [*Asarum europaeum* L.]

Hemerocallis Chalcedonica. [Probably *Hemerocallis fulva* (L.) L.]

Martagon, sive Lilium cruentum. [*Lilium martagon* L.]

Pulmonaria maculosa. [*Pulmonaria officinalis* L.]

Caryophyllus pratensis flore pleno, Bauh. Odontitis flore pleno. Cam. Clus. [Probably a form of *Dianthus plumarius* L.]

Iacea sive Flos trinitatis, Matth. Viola tricolor, Dod. Clus. [A form of *Viola tricolor* L.]

These were noted. We went back to the inn and had dinner. Then on the advice of our leader we made our way to Quex[1] and, exploring the ground with eager curiosity, found the following plants not till now collected on our journey: they were:

Sphondylium hirsutum vulgare; Branca vrsina, Trag. Acanthus vulgaris, Fuch. [*Heracleum sphondylium* L.]

Scrophularia major, Lob. Ocimastrum alterum, Trag. [*Scrophularia nodosa* L.]

Hieracium foliis & facie Chondrilla, Lob. [*Crepis capillaris* (L.) Wallr.]

Laureola, Dod. Daphnoides, Lob. Tab. Thymælea, Trag. [*Daphne laureola* L., still common in Quex Park]

[1] The manor of Quex or Quickes in Birchington parish was then the seat of Sir Henry Crispe (Hasted, 10, 298). R.

(Facs. pp. 80–1)

Hypericon vulgare. [*Hypericum perforatum* L.]
Ascyron, sive Hypericon in dumetis, 2. Trag. [*Hypericum hirsutum* L.]
Centaurium minus vulgare. [*Centaurium erythraea* Rafn (*C. umbellatum* auct.)]
Xyris spatula fœtida, Offic. [*Iris foetidissima* L.]
Oxalis, sive Acetosa vulgaris. [*Rumex acetosa* L.]
Polygonum mas vulgare, Lob. [*Polygonum aviculare* agg.]
Lithospermum, sive milium solis, Offic. [*Lithospermum officinale* L.]
Carlina minor flore purp. Clus. Carduus acaulis, Lob. [*Cirsium acaulon* (L.) Scop.]
Geranium arvense, Tab. Cicutæ folio, Bauh. [*Erodium cicutarium* (L.) L' Hérit.]
Viorna vulgi, Lob. 'Ατραγένη Theophrasti, Clus. Clematis 3. Matth. [*Clematis vitalba* L.]
Ptarmica vulgaris, Lob. Pyrethrum silvestre, Dod. Tanacetum album, Trag. [*Achillea ptarmica* L.]
Solanum lignosum, sive Dulcamara, Dod. Vitis syl. Matth. [*Solanum dulcamara* L.]
Lupulus salictarius. [*Humulus lupulus* L.]
Sanicula vulgaris, Diapensia, Matth. [*Sanicula europaea* L.]
Phallaris, semen canariense vulgo Belgis & Anglis. [*Phalaris canariensis* L.]
Bryonia nigra, Sigillum Beatæ Mariæ Offic. Tamus, Dod. [*Tamus communis* L.]
Bonus Henricus, Trag. Matth. Rumicis 3. Genus, Fuch. [*Chenopodium bonus-henricus* L.]
Cynocrambe, sive Mercurialis sylvestris, mas & fœm. [*Mercurialis perennis* L.]
Fragaria vulgaris. [*Fragaria vesca* L.]
Sison, Cordi, Petroselinum Macedonicum, Fuch. Dod. Amomum Germanicum, Tragi. [*Sison amomum* L.]
Acer majus, Lob. Platanus Scotica. Cardani. [*Acer pseudoplatanus* L.]
Acer minus vulgare. [*Acer campestre* L.]
Lychnis silvestris flore rubello. [*Silene dioica* (L.) Clairv. (*Melandrium rubrum* (Weigel) Garcke)]
Rhamnus Catharticus, sive spina infectoria, Matth. & Lob. [*Rhamnus catharticus* L.]
Filipendula, Saxifraga rubra, Tab. Oenanthe, Lob. Fuch. [*Filipendula vulgaris* Moench (*F. hexapetala* Gilib.)]
Prassium, Marrubium album. [*Marrubium vulgare* L.]
Chelidonium majus. [*Chelidonium majus* L.]

(Facs. p. 81)

Lysimachia siliquosa media. [Fl.K. gives *Epilobium obscurum* Schreb. as most likely; could be *E. tetragonum* L., or even a small *E. montanum* L.]

Argemone capitulo torulis canulato, sive Argemone Lacuna, Lob. [*Papaver hybridum* L.]

Argemone capitulo longiore, Ejusdem. [*Papaver argemone* L.]

Polygonum selinoides, Ger. sive Knawel Germanorum; an Vermiculata nova planta montana, Col? [Probably *Scleranthus annuus* L.]

Acinos parva flore cæruleo, sive Acinos Anglica, Clusii. [*Acinos arvensis* (Lam.) Dandy]

Pedicularis pratensis Lutea, sive Crista Galli herbariorum, Lob. [*Rhinanthus minor* L.]

Rubus saxatilis, Clusii. [Probably *Rubus idaeus* L. (see Fl.K.)]

Barbarea, sive pseudobunias, Lob. [*Barbarea vulgaris* R.Br.]

Sonchus asper vulgaris. [*Sonchus asper* (L.) Hill]

Sonchus Lævis tenerior, Lob. foliis minus dissectis. [*Sonchus oleraceus* L.]

Sonchus Lævis vulgaris foliis laciniosis, Ejusd. [*Sonchus oleraceus* L. (form with dissected leaves)]

Hipposelinum vulgare, Olusatrum, Petroselinum Alexandrinum, Trag. [*Smyrnium olusatrum* L.]

Blattaria flo. luteis vulgaris. [Probably *Verbascum virgatum* Stokes, but possibly the yellow form of *V. blattaria* L.]

Alsine corniculata, Clusii. [*Agrostemma githago* L., see Johnson's Gerard, p. 616]

[*Note.* In the margin against the following eleven plants is printed the note: 'These were growing in a small inlet commonly called Westgate Bay.']

Absynthium marinum sive Seriphium. [*Artemisia maritima* L.]

Alsine cruciata marina; vel Alsine pelagicum, Clusii. Anthillis prior lentifolia peplios effigie maritima, Lob. Ad. [*Honkenya peploides* (L.) Ehrh.]

Kali majus sive Salicornia. [*Salicornia* sp., probably an annual species.]

Kali minus. [*Suaeda maritima* (L.) Dum.]

Soldanella, κράμβη θαλασσία Dios. [*Calystegia soldanella* (L.) R.Br.]

Brassica marina, Monospermos, Lob. [*Crambe maritima* L.]

Eruca marina, sinapi marinum Ægyptium, Alpini. [*Cakile maritima* Scop.]

Serpentina major, Matthioli. [*Plantago coronopus* L. (possibly a large variety, e.g. var. *ceratophyllon* Rapin)]

Eryngium marinum vulgare. [*Eryngium maritimum* L.]

Glaux exigua maritima. [*Glaux maritima* L.]

(Facs. pp. 81–2)

Spergula marina, Ludg. Forte Anthylloides, Thal. ad salinas Saxonicas, & Alsine maritima Neapolitana, Col. [Probably *Spergularia marina* (L.) Griseb., more likely on a beach than *S. media* (L.) C. Presl]

[*Note.* The following three plants were no doubt found on the return from Westgate Bay to Margate. F.R.]

Cyperus rotundus inodorus septentrionalium, Lob. [*Scirpus maritimus* L.]

Paludapium, sive Apium, Offic. abundant everywhere in marshes and salt-marshes. [*Apium graveolens* L.]

Beta silvestris spontanea maritima, Lob. [*Beta vulgaris* L., subsp. *maritima* (L.) Thell.]

The whole day was spent in discovering these: and by sunset we were tired out and went home. Next morning, refreshed with food and sleep, we left Margate behind us and went on towards Sandwich. Before we got to the seashore, the following plants were collected:

[*Note.* On the way from Margate to Sandwich their route would lead by Pegwell Bay, and the former shingle beach of Stonar, now quarried away. F.R.]

Euphrasia vulgaris. [Almost certainly *Euphrasia nemorosa* (Pers.) Wallr.]

Sideritis alsines trixaginis foliis, Bauh. [*Stachys arvensis* (L.) L.]

Oenanthe Angustifolia, Lob. [Probably *Oenanthe lachenalii* C. C. Gmel., still common here.]

Cicutaria palustris, Lob. Phellandrium, Plinii. Dod. [*Oenanthe aquatica* (L.) Poir.]

Sagittaria aquatica, Plinii, Major Matth. Dod. Phleos mas latifol. Lugd. [*Sagittaria sagittifolia* L., not here now]

Hydrolapathum minus, Lob. [Probably *Rumex palustris* Sm., not here now]

Anagallis aquatica major foliis acutioribus, floribus albidis. [*Veronica anagallis-aquatica* L., rather than *V. catenata* Pennell]

Polygonon fœmina semine vidua, Lob. [*Hippuris vulgaris* L.]

Equisetum, 1. Matth. Hippuris major, Dod. [The name indicates *Equisetum telmateia* Ehrh., but the habitat makes *E. fluviatile* L. more likely]

Arundo vulgaris vallatoria, Lob. palustris, Matth. [*Phragmites communis* Trin.]

Hydropiper, persicaria acris. [*Polygonum hydropiper* L.]

Persicaria mitis maculosa. [Probably *Polygonum amphibium* L., but possibly *P. persicaria* L. or *P. lapathifolium* L.]

Iacea Floribus albis. [*Centaurea nigra* L., with white flowers]

(Facs. p. 82)

Gram. spica Tritici mutici, Bauh. [Fl.K. suggests *Agropyron junceiforme* (A. & D. Löve) A. & D. Löve, but this is unlikely, as the list was apparently not made on the shore; *A. pungens* (Pers.) Roem. & Schult., or its hybrid with *A. repens* (L.) Beauv., seems more likely]

Cichoreum sylvestre flore cæruleo. [*Cichorium intybus* L.]

Tragopogon vulgare luteo flore. [*Tragopogon pratensis* L.]

Cynoglossum majus. [*Cynoglossum officinale* L.]

Bardana, Lappa major. [*Arctium lappa* L.]

Gallium album, Tab. palustre, Dod. [*Galium palustre* L.]

Chamæpitys, sive Iva arthritica, Offic. [*Ajuga chamaepitys* (L.) Schreb.

Carlina silvestris, Dod. vulgaris, Clus. Atractylis, sive Cartamus syl. Trag. Carduus vulgaris, Matth. Hieracantha, Tab. [*Carlina vulgaris* L.]

Solanum lethale, Dod. mandragora, Theophrasti, Bella donna; Italorum. [*Atropa bella-donna* L.]

Eupatorium cannabinum mas. Herba St. Kunigundis, Trag. Eupatorium vulgare, Matth. Dod. Eupatorium Avicennæ vulgo creditum. [*Eupatorium cannabinum* L.]

Lysimachia siliquosa maxima hirsuta. [*Epilobium hirsutum* L.]

Lagopus, sive pes leporis. [*Trifolium arvense* L.]

Ladanum segetum, Lugd. flore albo. [*Galeopsis angustifolia* Hoffm., with white flowers]

[*Note*. The plants noted so far in this list include some from waste ground, possibly in the older shingle of Stonar Beach (e.g. *Galeopsis angustifolia*); a similar list could be made on the Rye Harbour shingles today. Some marshy pits (natural, or dug in the shingle for ballast) appear to have occurred, or else the marsh and aquatic species grew in dikes at Ebbsfleet. This fits in with the seventeenth-century topography about Pegwell and Stonar. Where *Ajuga chamaepitys* grew is uncertain—probably in the chalk at Pegwell among the corn. F.R.]

[*Note*. In the margin against the next four species is printed: By the sea shore.]

Caryophyllus marinus, sive gramen marinum. [*Armeria maritima* (Mill.) Willd.]

Iuncus major durior. [Fl.K. interprets this as *Juncus maritimus* Lam., which is still here, and Johnson himself (*Mercurius*, II. 24) names *J. acutus* L. 'Juncus maritimus acutus...(Pricking Sea Rush)'. Nevertheless, *J. acutus* L. has been known at Pegwell Bay, on Johnson's route, since at least 1829, and to the north of Deal since the time of Petiver and Sherard

(Facs. pp. 82–3)

—and is still abundant in both places. At Ebbsfleet it occurs in a relict dune-slack formed at a much earlier period than the present Sandwich dunes (i.e. pre-eighteenth century at least) and is a long-lived plant, slow to spread and unlikely to be a recent arrival. *J. acutus* L. was thus almost certainly seen by Johnson in 1632. It seems likely, therefore, that Johnson's 'Juncus major durior' (which fits *acutus* much better than *maritimus*) was, at least in part, *J. acutus* L.—making this the first British record]

Gramen sparteum 2. Schænanthinum, Tab. Spartum nostras, Lob. Spartum. 3. Clus. [*Ammophila arenaria* (L.) Link]

Lychnis marina Anglica, Lob. [*Silene maritima* With. Fl.K. gives Caryophyllus marinus as first records for both this species and *Armeria maritima* L.]

Then at last, crossing the river, we left Thanet and came to Sandwich. We went to an inn and stayed there a short time. Then two of us went off to the seashore towards Sandown Castle;[1] the others got ready to explore the town. Under the guidance of Mr Sparkes, a schoolmaster, they walked round the walls and the bastions, now partly ruinous with age, and entered the garden of Caspar Niren, a Belgian,[2] as also the apothecary's shop of Charles Duck, whom we afterwards met in Canterbury. In this place they saw a thing worth remembering, the 'spoils' (if I may so call them) of a serpent[3] fifteen feet long and thicker than an arm. As far as I can hazard a guess, it was a sea serpent; for it was caught by two men among the sandhills near the seashore, after its head had been shattered by small shot discharged from a fowling piece. It was hunting the rabbits, of which there is a vast abundance there, for food; for one or two were extracted from its stomach. These men, as I have said, brought the dead beast to our good friend Charles Duck, were duly rewarded and handed it over; its skin stripped from the flesh and stuffed with hay he still keeps with him as a memento of the event.

From Niren's garden, the seashore and the neighbourhood we got the following:

Malva vulgaris floribus albis. [*Malva sylvestris* L., with white flowers]

Anagallis aquatica, 3. Lobel. [*Samolus valerandi* L.]

Cotyledon aquatica, seu acris septentrionalium, Lob. [*Hydrocotyle vulgaris* L.]

[1] Sandown Castle north of Deal was built by Henry VIII in 1539; it was falling into ruin when Colonel Hutchinson the regicide died there in 1664 (Hasted, **10**, 13, 146). R.

[2] Though there was no such country as Belgium in 1632, the ancient name of the *Belgae* had been revived by classicists to denote the inhabitants of the southern Netherlands. L.

[3] See Appendix II, p. 145.

(Facs. p. 83)

Buglossum silvestre, Dod. [*Anchusa arvensis* (L.) Bieb. (*Lycopsis arvensis* L.)]
Tithymalus paralius. [*Euphorbia paralias* L.]

[*Note.* In the margin against the following ten animals is printed: In the sea.]

Ossa sepiae. [Cuttlefish 'bones' (*Sepia* sp.)]
Stella marina. [Starfish (*Asterias* sp.)]
Solen fœmina. [Both Pliny and Rondelet distinguished between male and female *Solen*! This is almost certainly the razor-shell, *Ensis ensis* (L.)]
Alcyonii duæ species. [The genus *Alcyonium* contains the deadman's fingers, *Alcyonium digitatum* (L.). Johnson may mean two colour forms of this species, but there are two other species of *Alcyonium* recorded for Britain]
Pectunculus, Rond. [A small species of *Chlamys* (a scallop); Rondelet figures this]
Echinus marinus sine aculeis minor, Besleri. [Probably the purple-tipped sea-urchin, *Psammechinus miliaris* (Gmelin), which has spines of nearly equal size. It occurs on the shore]
Echinus marinus, Spatagus, Rond, but so delicate and fragile that it could scarcely be handled. [*Spatangus purpureus* O. Fr. Müller (an irregular sea-urchin) or the heart-urchin *Echinocardium cordatum* (Pennant)]
Umbilici parvi. Boussueti. [The top-shell, *Gibbula umbilicalis* (da Costa)]
Concha longa latera Ejusdem. [Presumably a mollusc]
Lepades variæ magnitudinis. [The genus *Lepas* includes the goose-barnacles and this may be the correct interpretation, because goose-barnacles are often washed ashore. On the other hand *lepas* (Latin) can mean a limpet]

[*Note.* In the margin against the following eight species is printed: 'In Niren's garden.']

Myrrhis, Matth. Dod. Cerefolium Hispanicum. Tab. [*Myrrhis odorata* Scop.]
Glycyrhiza vulgaris. [*Glycyrrhiza glabra* L.]
Mercurialis mas & fœmina. [*Mercurialis annua* L.]
Valeriana græca, Dod. [*Polemonium caeruleum* L.]
Ageratum, Diosc. Eupatoreum Mesue, vulgo. [Probably *Achillea ageratum* L.]
Bistorta, Offic. serpentaria mas, Fuch. [*Polygonum bistorta* L.]
Caryophyllata montana, Matt. [*Geum montanum* L.]
Sambucus rosea. [*Viburnum opulus* L. 'Roseum']

(Facs. p. 84)

As night fell we returned to a supper seasoned not so much with variety of dishes as with harmless jests; and so, having relieved the dulness of our past journeying, we retired to rest. Next day, after paying our dues for our lodging and others for the stable, we left Sandwich. We had planned to go to Canterbury and setting forth on our way collected the following:

[*Note*. The following list relates to the area west of Sandwich, towards Canterbury. Much the same flora still occurs in the dikes west of the town. F.R.]

Morsus ranæ, Dod. Lob. [*Hydrocharis morsus-ranae* L.]

Iuncus floridus, Gladiolus aquaticus, seu palustris, Cordi. [*Butomus umbellatus* L.]

Potomogeiton angustifolium. [*Potamogeton* sp., or possibly *Polygonum amphibium* L.; the figure and description in Gerard are difficult to interpret]

Ranunculus tricophyllon aquaticus medioluteus, Col, sive Millefolium maratriphyllon tertium, &c. Lob. [Almost certainly *Ranunculus trichophyllus* sens. lat.]

Viola aquatilis, Dod. Palustris. Ger. [*Hottonia palustris* L.]

Ranunculus aquaticus hepaticæ facie, Lob. Polyanthemum aquat. Dod. [Almost certainly *Ranunculus peltatus* Schrank]

Raphanus aquaticus, Tab. [*Rorippa amphibia* (L.) Besser, rather than *R. islandica* (Oeder) Borbás]

Sium umbellatum repens, Matthioli & Italorum, as I think. [*Apium nodiflorum* (L.) Lag.]

Sium Crateva erucæfolium; Lob. sive Nasturtium aquatic. Offic. [*Rorippa nasturtium-aquaticum* (L.) Hayek (*Nasturtium officinale* R.Br.)]

Armeria pratensis, Lob. Caryophyllus pratensis, Tab. Flos cuculi pratensis, Trag. [*Lychnis flos-cuculi* L.]

Lens palustris. [*Lemna minor* L., and possibly other species of *Lemna*]

Sparganium ramosum, Platanaria, Dod. [*Sparganium erectum* L. var. *erectum* (*S. ramosum* Huds.)]

Typha major [*Typha latifolia* L.]

Thalictrum, sive Thalietrum, Ruta pratensis, Cordi. [*Thalictrum flavum* L.]

Œnanthe aquatica minor, Iuncus odoratus aquatilis, Dod. [*Oenanthe fistulosa* L.]

Lysimachia purpurea spicata, Lob. Solidaginis saracenicæ alia species, Trag. [*Lythrum salicaria* L.]

Buglossum echioides luteum, Lob. [*Picris echioides* L.]

(Facs. pp. 84–5)

Calamentha aquatica, Belgarum. [Fl.K. gives *Mentha arvensis* L., and Gerard's figure appears to be this, but *M. aquatica* L. is also possible]

Gramen palustre Cyperoides, Lob. [Probably *Glyceria maxima* (Hartm.) Holmberg]

Dipsacus vulgaris. [*Dipsacus fullonum* L. subsp. *fullonum* (*D. sylvestris* Huds.)]

Carduus lanceolatus, Tab. [*Cirsium vulgare* (Savi) Ten.]

Carduus sylvestris primus, Dod. [Probably *Cirsium palustre* (L.) Scop.]

Carduus sylvestris 3. Ejusd. sive Polyacantha Theophrasti, Lob. [Probably *Carduus crispus* L.]

Marrubium aquaticum, Sideritis prima, Matth. [*Stachys palustris* L.]

Betonica aquatica. [*Scrophularia aquatica* L.]

Filix mas. [*Dryopteris filix-mas* (L.) Schott]

Filix fæmina. [*Pteridium aquilinum* (L.) Kuhn]

[*Note*. About here the marshes are left, and the village of Ash reached. F.R.]

Adianthum album Plinii, Lob. [*Asplenium ruta-muraria* L., probably on the wall of Ash Church]

Genista vulgaris. [*Sarothamnus scoparius* (L.) Wimmer]

Saginæ spergula, Lob. Anthylloides, Thal. Alsine terrestris altera, τριχόφυλλος, Col. [*Spergula arvensis* L.]

Spergula flore rubro; an Alsine spergulæ facie minor, Bauh. Saxifraga Antiquorum. Adver. [*Spergularia rubra* (L.) J. & C. Presl]

Ranunculus arvorum semine echinato. [*Ranunculus arvensis* L.]

Lychnis silvestris parva. [The name, and the previous three species (indicating sandy arable land), suggest *Silene noctiflora* L., still in this part of Kent; if so, the first record for Britain]

Cracca major. Aracus, Lob. [Probably *Vicia sativa* L. subsp. *angustifolia* (L.) Gaud. on this Thanet sand soil]

Festuca altera, Dod. Lolium. 1. Trag. Bromos sterilis altera, Lob. [*Bromus* sp., possibly *B. secalinus* L.]

Thlaspi vulgare, sive vaccariæ folio, Lob. [*Lepidium campestre* (L.) R.Br.][1]

Lolium sive Triticum temulentum, Lob. [*Lolium temulentum* L.]

Carduus vulgatissimus radice repente, κεάνωθος Theophrasti, Col. [*Cirsium arvense* (L.) Scop.]

Gramen milliaceum. [Almost certainly not *Milium effusum* L. (which Johnson calls 'Gramen sylvaticum miliaceum' in the *Mercurius*), as the habitat (open sandy fields) is quite wrong. Possibly *Digitaria* sp. or *Panicum* sp.]

[1] See p. 51.

(Facs. p. 85)

Tussilago farfara, Vngula Caballina. [*Tussilago farfara* L.]

Symphytum majus, sive Consolida major vulgaris. [*Symphytum officinale* L.]

Petasites, sive Tussilago major, Matth. [*Petasites hybridus* (L.) Gaertn., Mey. & Scherb., still in ditches at Preston, near Wingham.]

[*Note*. At this point, woodland seems to be reached. F.R.]

Hypericum minus repens. [*Hypericum humifusum* L.]

Veronica minor serpillifolia, Lob. [*Veronica serpyllifolia* L.]

Aracus, Tab. Galegæ silvestris, Dod. [Johnson's *Vicia* spp. are difficult to interpret; this is probably either *V. sepium* L. or *V. sativa* L.]

Carduus mariæ, Leucographis Plinii. [*Silybum marianum* (L.) Gaertn.]

Linum silvestre perpusillum flo. albis non descriptum. [*Linum catharticum* L.]

Vicia, Tab. Legumen terræ glandibus simile, Dod. Aphacoides. [*Lathyrus pratensis* L.]

Caltha palustris, Tussilago altera, Matth. [*Caltha palustris* L.]

Acorus nostras palustris, Lob. Iris palustris lutea, Tab. Pseudoiris, Dod. [*Iris pseudacorus* L.]

Iuncus capitulis sparsis. [*Juncus effusus* L.]

Trifolium corniculatum majus hirsutum. [*Lotus pedunculatus* Cav. (*L. uliginosus* Schkuhr)]

Anchusa degener facie milii solis, Lob. Anchusa arvensis minor, Tab. Lithospermum silvestre, Trag. [*Lithospermum arvense* L.]

Anonis sive Ononis spinosa. [*Ononis spinosa* L.]

Anagallis aquat. vulgaris foliis rotundioribus floribus cæruleis, sive Becabunga, Offic. [*Veronica beccabunga* L.]

Gramen aquis innatans, Lob. [*Glyceria fluitans* (L.) R.Br., or perhaps *G. plicata* Fr.]

Ranunculus hederaceus, Lugd. [*Ranunculus hederaceus* L., or a related species.]

Ranunculus aquaticus rotundifolius, forte Apium risus, Lob. [Either *Ranunculus hederaceus* L., or a related species]

Helenium, sive Enula Campana, Offic. [*Inula helenium* L.]

Sedum majus vulgare, Cotyledon altera prima, Clus. [*Sempervivum tectorum* L.]

Sedum medium teretifolium, Lob. Sedum minus alterum, Clus. [Either *Sedum album* L. or S. *dasyphyllum* L., more probably the latter]

Vlmaria, Regina Prati, Dod. Barbicapra, Lob. [*Filipendula ulmaria* (L.) Maxim.]

(Facs. pp. 85–6)

Alsine aquatica, Lugd. stellaria, Lob. [*Stellaria alsine* Grimm. (*S. uliginosa* Murr.)]

Hedera terrestris, Chamæcissos. [*Glechoma hederacea* L.]

Arum vulgare. [*Arum maculatum* L.]

Alsine maxima. [Probably *Myosoton aquaticum* (L.) Moench, or possibly *Stellaria neglecta* Weihe]

Gramen tomentosum, & acerosum, Calamogrostis quorundam, Lob. [*Calamagrostis epigejos* (L.) Roth]

Agrorum venti spica, Lob. [*Apera spica-venti* (L.) Beauv.]

Cannabis sativa fœmina. [*Cannabis sativa* L.]

Malva pumila, seu syl. repens flore parvo albido. [*Malva neglecta* Wallr.]

[*Note*. The following list could be made today—in nearly all its components—in the woods east of Canterbury (Oldchurch, Pine, and Trenley Park Woods). An asterisk marks those that I have seen myself. F.R.]

Gnaphalium Anglicum, vel Belgicum folio longiore, Lob. [**Gnaphalium sylvaticum* L.]

Filix saxatilis, Clus. [*Asplenium adiantum-nigrum* L.]

Taxus, Σμίλαξ, Dios. Μίλος, Theoph. [**Taxus baccata* L.]

Bryonia alba, Dod. Vitis alba, Matth. [**Bryonia dioica* Jacq.]

Cannabis silvestris spuria, Lamium quorundam flo. purp. [**Galeopsis tetrahit* L.]

Pusillum fontilapathum, Lob. Tribulus aquaticus minor, Clus. [*Potamogeton crispus* L.]

Ranunculus flammeus foliis non serratis minor. [**Ranunculus flammula* L.]

Ranunculus flammeus foliis serratis. [A form of **Ranunculus flammula* L.]

Cratæogonon, Lob. Milium sylvaticum, Tab. Sideritis pratensis lutea, Lugd. [**Melampyrum pratense* L.]

Betonica vulgaris. [**Betonica officinalis* L.]

Trachelium majus sive Cervicaria, Lob. Uvularia major, Trag. [**Campanula trachelium* L.; seen on Thanet sand (calcareous) in 1949.]

Aracus, sive Cracca minima, Lob. [**Vicia hirsuta* (L.) S. F. Gray most likely, but * *V. tetrasperma* (L.) Schreb. possible.]

Gnaphalium minimum, Lob. [*Filago minima* (Sm.) Pers. according to Fl.K., but possibly *Gnaphalium uliginosum* L.]

Millegrana minima, Lob. polygonum polyspermum, quorundam. [*Radiola linoides* Roth]

(Facs. p. 86)

Tithymalus charachias amygdaloides. [**Euphorbia amygdaloides* L.]

Numularia, Matth. Dod. Lob. Hirundinaria minor, Tab. [**Lysimachia nummularia* L.]

Armeria silvestris altera calyculo foliolis fastigiatis cincto, Lob. [**Dianthus armeria* L.; still near the main road in 1963]

Ranunculus flammeus major, Tab. Ranunculus πλατύφυλλος, Thal. [The name refers to *Ranunculus lingua* L., but more likely a robust form of **R. flammula* L.]

Tormentilla vulgaris, etiam consolida rubra, Tab. Heptaphyllum, Fuch. [**Potentilla erecta* (L.) Räusch.]

Vicia maxima dumetorum, Bauh. Cracca major, Tab. [Probably **Vicia cracca* L., rather than *V. angustifolia* L.]

Angelica silvestris. [**Angelica sylvestris* L.]

Alnus, Matth. Dod. [**Alnus glutinosa* (L.) Gaertn.]

Sambucus aquatica, Matth. Lob. Trag. palustris, Dod. [**Viburnum opulus* L.]

Bellis major, Matth. Dod. Trag. Consolida media vulnerariorum. Lob. [**Chrysanthemum leucanthemum* L.]

Trifolium acetosum, Lujula, sive Alleluia, Offic. [**Oxalis acetosella* L.]

Adianthum aureum majus, Tab. Polytrichum majus, Trag. Fuch. [Probably *Polytrichum formosum* Hedw., or possibly *P. commune* Hedw.] See Appendix III, p. 147.

Adianthum aureum minus, polytrichum minus, Trag. Fuch. See Appendix III, p. 148.

Musci capillaris altera species, Dod. [Possibly *Funaria hygrometrica* Hedw.] See Appendix III, p. 150.

Leucojum Luteum, Dod. Keiri, Lob. & Offic. Viola petræa lutea, Tab. [**Cheiranthus cheiri* L.; still on the walls of Canterbury]

[*Note*. In the margin against the next two plants is printed: 'In a garden'.]

Flos solis Farnesianus, sive Aster Peruvianus, Col. [*Helianthus tuberosus* L.]

Sumach, sive Rhus Virginianum. [*Rhus typhina* L.]

As the sun was now setting in the west, we entered Canterbury and did not wait long before we visited the Cathedral Church of the Archbishopric of Canterbury, and there attended divine service celebrated in the customary way. When it was over, we explored the cathedral, once world-famous for the shrine of Thomas Becket, of which 'the least costly part was of gold'.[1]

[1] Camden's *Britannia*, c. 194. The shrine was destroyed in 1538 by Thomas Cromwell. R.

(Facs. pp. 86–7)

Then we visited the various tombs and above all and most eagerly that of the warrior prince Edward called the Black,[1] and were equally delighted with the wonderful artistry and taste of the stained glass windows. Then we climbed the towers and surveyed the city and its walls falling into ruin. For our people, like the Spartans of old, set more store upon arms than upon walls for protection.

Here we happened to meet Charles Duck, our very good friend as I have said, and were made welcome by the kindness of Richard Jackson and equally so by the friendliness of a man eminent for musical ability, William Pysing,[2] who was not only with us while we were in Canterbury but accompanied us as far as London.

The Lord's day, as was appropriate, was dedicated to rest and divine worship. Next morning we bent our energies again to our usual work, and the following plants were found:

[*Note*. This list covers Canterbury to Faversham, probably including Blean Woods by Watling Street. F.R.]

Teucrium pratense, Lob. Verbenaca supina mas. Dod. Chamædrys silvestris, Clus. Ger. [*Veronica chamaedrys* L.]

Hieracium longius radicatum. [*Hypochoeris radicata* L.]

Hieracium montanum angustifolium primum, Tab. [Probably *Hieracium umbellatum* L.]

Hieracium fruticosum latifolium glabrum, Bauh. [*Hieracium* sp., probably *H. perpropinquum* (Zahn) Druce or *H. rigens* Jord.]

Veronica pratensis, Dod. Veronica minor, Tab. [*Veronica serpyllifolia* L.]

Hypericum pulchrum, Tragi. [*Hypericum pulchrum* L.]

Digitalis purpurea, Dod. Lob. Ephemeri Dios. alia species major, Col. Campanula syl. Trag. [*Digitalis purpurea* L.]

Polygala flo. cæruleo, & flo. albo, Amarella Gesn. flos Ambervalis Dod. [Almost certainly *Polygala serpyllifolia* Hose, not *P. vulgaris* L., in the acid Blean Woods]

Ruscus, Bruscus, Oxymyrsine, Myrtacantha, Lob. [*Ruscus aculeatus* L.]

Alsine aquatica foliis rotundioribus, sive Portulaca aquatica. [*Peplis portula* L. See Fig. 4, p. 96.]

Argentina, Dod. Lob. Potentilla, Matth. Anserina, Trag, Tab. [*Potentilla anserina* L.]

[1] Camden, *loc. cit.* The Prince had died in the Archbishop's Palace there in 1376 (Stow, *Annals*, ed. 1615, p. 271). R.

[2] Named in the Treasurer's Accounts for the Cathedral (nos. 39–44) for 1631–7 as a lay clerk who could, when required, act as a minor canon. He was Master of the Choristers 1639–63. R.

(Facs. pp. 87–8)

Serpillifolia aquatica, Bauh. sive Alsine palustris serpillifolia. Vide hujus fig. in fine lib. notatam, fig. 5. [*Callitriche* sp., probably *C. stagnalis* Scop.] See Fig. 5, p. 96.

Saxifraga minor altera, flosculis albis, sem. nigro, forte Alsine sassifraga angustifolia minima montana, Colum. Anglica saxifraga. Ad. [Possibly *Montia fontana* L.]

Cicutaria alba, Ludg. Cicutaria palustris, flore candido, Cam. [*Oenanthe aquatica* (L.) Poir.]

We saw nothing except these, and many noted before, between Canterbury and Faversham. Halting there, some of us set out towards the sea and gathered:

Limonium vulgare seu perperam, vt etiam Behen rubrum. [*Limonium vulgare* Mill.]

Peucedanum vulgare seu Fœniculum porcinum. [*Peucedanum officinale* L.]

Plantagio aquatica, Limonium verum Dios. vt alias ostendam. [*Alisma plantago-aquatica* L.]

Althæa vulgaris, sive Bismalva, Offic. & Ger. [*Althaea officinalis* L.]

Tripolium vulgare, Amelli species palust. Col. [*Aster tripolium* L.]

Allium silvestre tenuifolium, Lob. Ophioscoridon, Matth. [*Allium vineale* L.]

Then we went to the shop of Nicholas Swayton, a fellow of our craft.[1] He welcomed us kindly and took us into his garden, where the following herbs were growing:

Cucumer silvestris, Matth. Cucumer Asinius. Offic. Tab. [*Cucumis sativus* L.]

Aster atticus, sive Amellus, Matth. Tintorius, flos. 1. Trag. Aster Italorum purpurascens, Lob. [*Aster amellus* L.]

Leucoium purpureum. [Probably *Matthiola incana* (L.) R.Br.]

Chondrilla rubra fœtida, Eyst. [*Crepis rubra* L.]

Majorana. [*Origanum* sp., or related genus]

Marum vulgare, Mastic, Gallorum & Anglorum. Ad. Trag. Origanum Hispani cum primum, Clus. [*Origanum* sp., or related genus]

Aquilegia. [*Aquilegia vulgaris* L.]

Chamæcyparissus. [*Santolina chamaecyparissus* L.]

Asparagus vulgaris. [*Asparagus officinalis* L.]

Carduus benedictus. [*Cnicus benedictus* L.]

Cardiaca, Matth. Lob. Tab. Melissa syl. Tragi. [*Leonurus cardiaca* L.]

[1] He is mentioned as an honest and skilful apothecary in Johnson's edition of Gerard's *Herball*, p. 303. R.

(Facs. p. 88)

Cameline, Myagrum alterum Thlaspi effigie, Lob. [*Erysimum cheiranthoides* L.]

Papaver flore albo multipl. [*Papaver somniferum* L., double form]

Cyclamen vernum, panis porcinus, Rapum seu Umbilicus terræ, Quorund. [*Cyclamen* sp.]

Parthenium, matricaria flore multiplice. [*Chrysanthemum parthenium* (L.) Bernh.]

Caryophyllorum hortensium species variæ. [*Dianthus* spp.]

Polium luteum montanum, Tab. Ger. [Probably *Teucrium aureum* Schreb.]

Melilotus vulgaris Offic. [*Melilotus* sp., probably *M. altissima* Thuill.]

After we had seen these, he offered to be our guide and took us to a wood full of different plants: I omit many already mentioned; but there grew there:

Blattaria flo. albis, Lob. [*Verbascum blattaria* L.; still in Putt Wood, 2 miles S.W. of Faversham]

Androsæmum, Dod. Clymenon Italorum, sive Siciliana, Lob. [*Hypericum androsaemum* L.]

Holostium Ruellii, Gramen leucanthemum, Dod. [*Stellaria holostea* L.]

Cannabis sil. spuria, Lamium quorundam flo. albis. [*Galeopsis tetrahit* agg., probably *G. bifida* Boenn.]

Virga aurea, Matth. Dod. Symphytum petræum. 3. Tab. [*Solidago virgaurea* L.]

Primula veris, sive Primula sylvarum, Lob. Alisma sylvarum, Col. [*Primula vulgaris* Huds.]

Oxalis, sive Acetosa vervicina tenuifolia, Lob. Oxalis ovina, Tab. [*Rumex acetosella* agg.]

Gramen avenaceum rariore grano nemorense, Lob. [*Melica uniflora* Retz.]

Gramen nodosum avenacea pannicula. [*Arrhenatherum elatius* (L.) J. & C. Presl]

Gramen Phalaroides majus, Lob. [Possibly *Phalaris arundinacea* L.]

Agrifolium, Dod. Aquifolium, Matth. Lob. [*Ilex aquifolium* L.]

Viola canina, cærulea inodora, sylvestris, serotina, Lob. [Almost certainly *Viola riviniana* Rchb.]

Gramen caninum medicatum vulgare. [*Agropyron repens* (L.) Beauv.]

Bugula, Dod. Lob. Consolida media, Matt. Fuch. Tab. [*Ajuga reptans* L.]

Orchis myodes prima. [*Ophrys insectifera* L. (*O. muscifera* Huds.)]

Circæa Lutetiana, Lob. Lappa sylvestris, Trag. Herba D. Stephani, Tab. [*Circaea lutetiana* L.]

(Facs. pp. 88–9)

Scordium alterum Plinii, Lob. Salviagrestis, Dod. Sylvestris, Trag. [*Teucrium scorodonia* L.]

Polypodium. [*Polypodium vulgare* L.]

Glaux vulgaris, Lob. Glycyrhiza sylvestris, Gesn. Fænugræcum syl. Trag. [*Astragalus glycophyllos* L.; still south of Faversham]

Leaving the wood and thanking our friend, we took the road from Canterbury to London and with what speed we could came through Sittingbourne to Rochester and so at last to Gravesend. In addition to plants already listed, these attracted our notice:

Muscus aquaticus denticulatus. [*Fontinalis antipyretica* Hedw.] See Appendix III, p. 150.

Alopecuros maxima Anglica paludosa, Lob. [Probably *Polypogon monspeliensis* Desfont., as suggested in Fl.K. and other works. The locality was probably along the Swale estuary, or near Strood, or in Milton Creek]

Ruta muraria, Dod. Capillus Veneris, Trag. Salvia vitæ, Lob. Ad. Lugd. Ad, Lugd. Adianthum album. Tab. paronychia, Matth. [*Asplenium ruta-muraria* L.]

Calamentha pulegii odore, Lob. [*Calamintha nepeta* (L.) Savi; still plentiful at Key Street and Chestnut Street near Sittingbourne and, until recently, along Watling Street near Norton]

Quinquefolium peiræum majus, Tab. Tormentilla facie, Ger. Pentaphyllum album, Matth. exiguum alterum, Tragi. [*Potentilla reptans* L.]

Pilosella repens vulgaris. [*Hieracium pilosella* L.]

Phyllitis, Matth. Dod. Lingua cervina, Offic. Scolopendria vulgaris, Trag. [*Phyllitis scolopendrium* (L.) Newm.]

Aphaca, Lob. Dod. Elatine 3. Tab. [*Lathyrus aphaca* L.]

Antirrhinum medium. [*Antirrhinum orontium* L.]

Tanacetum vulgare, Matth. Dod. Artemisia Diosc. Tab. [*Tanacetum vulgare* L. (*Chrysanthemum vulgare* (L.) Bernh.]

Castanea vulgaris. [*Castanea sativa* Mill.]

Muscus terrestris pyxidatus alabastriculos imitatus, Lob. [*Cladonia pyxidata* (L.) Hoffm.] See Appendix III, p. 147.

Synanchica, Lugd. Gallium montanum cruciatum, Col. Rubiæ genus parvum, Ges. Col. [*Asperula cynanchica* L.]

Chamæcistus flore luteo. Helianthemum, Lob. Panax chironium, Matth. Hyssopus campestris, Trag. [*Helianthemum chamaecistus* Mill.]

Linum vulgare sativum. [*Linum usitatissimum* L.]

(Facs. pp. 89–90)

Lathyrus latiore folio, Lob. Clymenum, Matth. [*Lathyrus sylvestris* L.; still by the main road at Higham between Gravesend and Rochester]

Saponaria, Dod. Lob. Trag. Matth. Struthium, Fuch. [*Saponaria officinalis* L.]

Coniza cærulea acris; Tinctorius flos alter, Tragi. Amellus mont. Col. [*Erigeron acer* L.]

Then, after waiting for an opportune moment at Gravesend, at the first of the tide on the river we entered a splendid boat driven by eight oars that happened to be left there and so returned to London, giving heartfelt thanks to God for His many benefits conferred on us: and we pray that on the works undertaken by us and all men for the public good He will bestow the desired fulfilment. *Amen.*

THE END

(Facs. p. 90)

List of Plants Growing on Hampstead Heath and its Neighbourhood

I THINK it worth while to enumerate here the plants growing on Hampstead Heath and in the places visited by us on the way there and back and not included in the previous Itinerary. In my previous work I noted some plants that are common with us almost everywhere. Here I have thought fit to name others no less familiar. For to those untrained in this field and interested in medicine and herbal remedies the knowledge of these common plants and their names is no less needful than of others. I have purposely omitted a few trees—those grown for their fruit and known to almost all, such as apple, pear, cherry and the like. So also I have made no mention of cereals and cultivated vegetables, such as rye, wheat, barley, peas, beans, and the like. Other trees, however common, I have made a point of including. I have appended synonyms for certain less known plants where appropriate, and for those variously named by different authors.

And now to business. You will find the commonest and best known species first in the list:

Atriplex silvestris latifolia, Lob. silvestris, 3 Matth. Blitum. 4. Trag. pes anserinus, Dod. [*Chenopodium rubrum* L.]

Atriplex olida, Lob. Canina, Tragi. Garosmum, Dod. [*Chenopodium vulvaria* L.]

Alsine hederacea, Tab. Morsus gallinae folio hederulae, Lob. [*Veronica hederifolia* L.]

Alsine minor, Tab. minima, Dod. Lob. minor multicaulis, Bauh. [*Arenaria serpyllifolia* L., sens. lat.]

Eupatorium cannabinum fœmina, Lob. Hepatorium aquatile, Dod. Verbena supina, Trag. [*Bidens tripartita* L.]

Eruca sylvestris, Lob. [*Diplotaxis tenuifolia* (L.) DC.]

Sinapi sylvestre minus bursæ pastoris folio, Lob. Sinapi. 3. Matth. an Irio Apulus alter levifolio Erucæ. Col. [Probably also *Diplotaxis tenuifolia* (L.) DC.]

Myrrhis sylvestris nova, Æquicolorum, Col. [*Anthriscus caucalis* Bieb. (*A. vulgaris* Pers.)]

(Facs. pp. 90–1)

Paronychia vulgaris, Dod. alsinefolia, Lob. Bursa pastoria, 6. Trag. [*Erophila verna* (L.) Chevall.]

Paronychia altera, Dod. rutaceo folio, Lob. Alsine petræa rubra, & Paronychia, 3. Tab. [Possibly *Teesdalia nudicaulis* (L.) R.Br.]

Coniza minor, Trag. Matth. minima, Lob. Dod. [*Pulicaria vulgaris* Gaertn.]

Smilax lævis. Matth. Dod. Volubilis major, Tab. Trag. [*Calystegia sepium* (L.) R.Br. (*Convolvulus sepium* L.)]

Lactuca syl. odore viroso foliis dissectis, Lactuca sylv. prior, Trag. silv. Matth. [*Lactuca serriola* L. (*L. scariola* L.)]

Lactuca syl. alter odore magis viroso foliis non dissectis, Lactuca agrestis odore opii, Lob. Ad. Endivia major & 1. Trag. Thesion, Lugd. [*Lactuca virosa* L.]

Medica minor fructo cochleato aspero; Tribulus terrestris minor repens, Lugd. vt etiam ejus varietas foliis maculatis, Medica Arabica, Cam. [*Medicago arabica* (L.) Huds. (*M. maculata* Sibth.)]

Rosa canina sylvestris inodora, Lob. [Probably a form of *Rosa canina* agg.]

Rosa silvestris pomifera, Lob. Arvina, Tab. [*Rosa pimpinellifolia* L. (*R. spinosissima* L., p. p.)]

Valeriana silvestris major. [*Valeriana officinalis* L.]

Hyacinthus vulgaris, Anglicus & Belgicus, Lob. [*Endymion non-scriptus* (L.) Garcke.]

Alliaria, Matth. Dod. [*Alliaria petiolata* (Bieb.) Cavara & Grande.]

Potomogeiton majus vulgare, Matth. Trag. [*Potamogeton natans* L.; Fl.M. quotes Johnson, but gives 'Petiver, about 1710' as first record for the county!]

Saxifraga Anglicana facie Seseli pratensis, Lob. [*Silaum silaus* (L.) Schinz & Thell.]

Alectorolophos, pedicularis rubra, sive Ruta pedicularis, Tab. Fistularia. Dod. [*Pedicularis sylvatica* L.]

Ranunculus hederaceus aquaticus, Lugd. [*Ranunculus hederaceus* L.]

Gramen pratense vulgare. [Probably *Poa pratensis* L.]

Gramen loliaceum. [Fl.M. gives *Lolium temulentum* L.; but *L. perenne* L. seems more likely]

Gramen spicatum foliis vetonicæ caryophyllatæ, Lob. [*Carex caryophyllea* Latour. (*C. praecox* auct.)]

Gramen palustre echinatum, Lob. aculeatum, Lugd. [*Carex otrubae* Podp. (*C. vulpina* auct., non L.)]

Gramen tremulum, sive phalaris pratensis. [*Briza media* L.]

Gramen exile hirsutum cyperoides, Lob. [*Luzula campestris* (L.) DC.]

(Facs. p. 91)

Gramen spartium capillaceo folio minimum. [*Nardus stricta* L.]
Gramen holostium minimum, forte, Holostium Alpinum minimum, Bauh. [*Juncus bufonius* L.]
Gramen junceum sylvaticum, Tab. Ger. [*Juncus acutiflorus* Hoffm.]
Ranunculus bulbosus, Lob. tuberosus, Dod. [*Ranunculus bulbosus* L.]
Chelidonium minus, sive Scrophularia minor. [*Ranunculus ficaria* L.]
Fumaria vulgaris. [*Fumaria officinalis* L.]
Fraxinus. [*Fraxinus excelsior* L.]
Quercus. [*Quercus robur* L.]
Rubus vulgaris. [*Rubus fruticosus* L. agg.]
Sambucus. [*Sambucus nigra* L.]
Betula, Matth. Lob. [Probably *Betula pubescens* Ehrh.]
Ligustrum. [*Ligustrum vulgare* L.]
Corylus sylvestris, Matth. Nux avellana syl. Dod. [*Corylus avellana* L.]
Erica vulgaris, Trag. Ger. prima, Matth. Dod. [*Calluna vulgaris* (L.) Hull.]
Erica pumila calyculato vnedonis flore, Lob. [*Erica cinerea* L.]
Muscus terrestris vulgaris. [Possibly *Sphagnum* sp.] See Appendix III, p. 148.
Muscus in ericetis proveniens, Lob. [Possibly *Bryum capillare* Hedw.] See Appendix III, p. 149.
Muscus corallinus albicans, sive corallina, montana, Tab. [Probably *Cladonia sylvatica* Hoffm.] See Appendix III, p. 149.
Oxyacantha, Matth. Lob. Spina appendix Plinii. Gesn. [*Crataegus monogyna* Jacq., and possibly *C. oxyacanthoides* Thuill.]
Genestella, Trag. Dod. Genistella aculeata, Lob. [*Genista anglica* L.]
Genista Tinctoria, Dod. Genistella infectoria, Lob. [*Genista tinctoria* L.]
Genista spinosa vulgaris. [*Ulex europaeus* L.]
Iuniperus vulgaris fruticosa. [*Juniperus communis* L.]
Salix vulgaris longis & angustis foliis. [Probably *Salix viminalis* L.]
Salix latioribus albidis foliis. [Probably *Salix caprea* L.]
Salix humilis repens, Lob. bombicifera. [*Salix repens* L.]
Prunus sylvestris foliis latioribus, fructu majore. [*Prunus insititia* L.]
Prunus sylvestris, Matth. Dod. Lob. Acacia Germanica, Quorundam. [*Prunus spinosa* L.]
Populus Lybica, Matth. Dod. Cercis Theophrasti. Populus Lybica Plinii, Clus. [*Populus tremula* L.]
Carpinus, Matth. Ornus, Trag. Ostrys Theophrasti, Clus. Betulus, Lob. [*Carpinus betulus* L.]

(Facs. pp. 91–2)

Sorbus sylv. Alpina, Lob. Ornus, & fraxinus bubula, Dod. Fraxinea arbor Trag. [*Sorbus aucuparia* L.]

Sorbus torminalis Plinii, Matth. Lob. Dod. [*Sorbus torminalis* (L.) Crantz]

Alnus nigra baccifera, Lugd. Frangula, Matth. [*Frangula alnus* Mill.]

Aria Theophrasti effigie Alni, Lob. Sorbus Aria cognominata, Clus. [*Sorbus aria* (L.) Crantz]

Veronica vera & major, Lob. Veronica mas, Dod. Fuch. [*Veronica officinalis* L.]

Cynosorchis morio, Lob. Orchis mas angustifolia, Fuch. [*Orchis mascula* L.; Fl.M. quotes this record, but gives Blackstone 1737 as the 'first record' for the county!]

Orchis ornithophora, Lob. [Possibly *Orchis morio* L., see Johnson's Gerard, p. 215; if so, first record for Middlesex]

Serapias candido flore montana maculatis foliis, Lob. [Almost certainly *Dactylorhiza maculata* (L.) Soó subsp. *ericetorum* (Linton) Hunt & Summerh. (*Orchis ericetorum* (Linton) Marshall)]

Palma Christi, Serapias fæmina pratensis, Ejusd. [Probably also a species of *Dactylorhiza*]

Primula pratensis, Lob. Herba paralysis vulgaris, Trag. Alisma pratorum, Col. [*Primula veris* L.]

Bulbocastanum, Dod. Nucula terrestris Septentrionalium, Lob. [*Conopodium majus* (Gouan) Loret (*C. denudatum* Koch)]

Chamæmelum, sive Anthemis vulgatior, Lob. [*Matricaria recutita* L. (*M. chamomilla* auct.)]

Cardamine altera Lob. flos Cuculi, Dod. Sisymbrium aquaticum alterum, Matth. Tab. [*Cardamine pratensis* L.]

Cyperus gramineus, sive miliaceus, Lob. [*Scirpus sylvaticus* L.]

Cruciata, Dod. Minor, Lob. Gallium. 2. Trag. [*Galium cruciata* (L.) Scop.]

Alcea vulgaris, Dod. Matth. [*Malva moschata* L.]

Lameum luteum, Lob. Tab. [*Galeobdolon luteum* Huds. (*Lamium galeobdolon* (L.) L.)]

Ervum sylvestre, Dod. Silvestre, herbariorum, Lob. & Catanance, Quorundam. [*Lathyrus nissolia* L.]

Succisa, Matth. Dod. Morsus diaboli, Trag. Lob. [*Succisa pratensis* Moench (*Scabiosa succisa* L.)]

Bifolium, Lob. Ophris, Matth. [*Listera ovata* (L.) R.Br.]

Filix florida, vel Osmunda regalis, Lob. filix aquatica, Dod. [*Osmunda regalis* L.]

(Facs. pp. 92–3)

Lonchitis altera, Dioscoridis, Lob. Asplenium silvestre, Trag. Lonchitis aspera minor, Matth. [*Blechnum spicant* (L.) Roth]

Vaccinia nigra, Dod. Myrtillus, Matth. Vitis Idaea vulgaris. Clus. Idaea. 1. Tab. [*Vaccinium myrtillus* L.]

Iuncus bombycinus, Lob. Linum pratense. Ges. Gnaphalium, Trag. Gramen tomentarium, Ger. [*Eriophorum angustifolium* Honck.]

Trifolium paludosum, Lob. Ger. Limonium, Cordi. Menianthes palustre, Lugd. [*Menyanthes trifoliata* L.]

Ros solis, Dod. Rorella prima, Tab. [*Drosera rotundifolia* L.]

Muscus terrestris, Matth. Dod. Muscus clavatus, Lob. Ger. Lycopodium, Tab. [*Lycopodium clavatum* L.]

Orobanche, Rapum genistæ. [*Orobanche rapum-genistae* Thuill.]

Lilium convallium, Trag. Matth. [*Convallaria majalis* L.]

Scabiosa montana minima, Lob. [*Jasione montana* L. This record is not mentioned in Fl.M., which gives 1666 as first record]

Lysimachia galericulata, Lob. Sideritis cærulea, Thal. Tertianaria, Tab. Gratiola, latifolia, & Cærulea, Quorundam. [*Scutellaria galericulata* L.]

Lysimachia galericulata minor, sive Gratiola latifolia, Ger. [*Scutellaria minor* L.]

Ornithopodium perpusillum, Lob. [*Ornithopus perpusillus* L.]

Anagallis flore luteo, Ejusd. [*Lysimachia nemorum* L.]

Serratula, Matth. Dod. Centauroides, vel centaurium majus sylvestre Germanicum, Thal. [*Serratula tinctoria* L.]

Astragalus sylvaticus, Thal. [*Lathyris montanus* Bernh. (*L. macrorrhizus* Wimm.)]

Cirsium Anglicum, Lob. [*Cirsium dissectum* (L.) Hill (*Carduus pratensis* Huds.)]

THE END

(Facs. p. 93)

Explanation of the Illustrations[1]

1.2. Fucus longiss. latissimo, crassoq; folio. Bauh. etc. [Fig. 1 is *Laminaria saccharina* (L.) Lamour, and Fig. 2 *L. digitata* (Huds.) Lamour. See Appendix 1, p. 140.] It is strikingly variable. Sometimes there springs from a fibrous root and a pedicel half a foot long a single leaf a full ell in length; in that case it is almost a hand's breadth wide, so that it nicely represents a two-edged sword. Sometimes it puts out two leaves like that just described except in length.

Sometimes it puts out a polyschides or much-divided leaf, which is divided into eight, nine, ten, or twelve parts more or less. Sometimes one undivided leaf and another much-divided come up from the same root, as is shown in the figure attached. So in that case the part numbered 1 is known as Fucus Phasganoides [sword-like] and the other marked 2 Polyschides.

So much for the shape. Now for the colour. This is not the same in all cases. For some are more green, and these can hardly be dried. Others are whitish and dry at once: they look very like parchment in colour and consistency, so that they can easily be taken for it by those who do not know this. The first variety of it is well described while still growing by Bauhin in his Prodromus, Book 10, Chapter 8, numbers 1. 2. and especially 3. under the name of Fucus Longiss. etc.

3. Fucus spongiosus nodosus, in English Sea Ragged-staffe. [See Appendix 1, p. 141.]

This is a very succulent and fungus-like plant of the thickness of a thumb. It is of a dull yellow, and swells out in many uneven appendages or knots, whence it was aptly called by our Warden, Thomas Hicks, in English, Sea Ragged-staffe. We did not see it growing, but found one or two plants a foot long.

4. Alsine aquatica foliis rotundioribus, sive Portulaca Aquatica. [*Peplis portula* L.]

This is described by Caspar Bauhin in the Prodromus, p. 118, under the name Alsines palustris minoris folio oblongo.

5. Serpillifolia aquatica Io. Bauhini, etc. [*Callitriche* sp., probably *C. stagnalis* Scop.]

[1] See plate in the Facsimile, p. 96.

(Facs. p. 94)

It is described along with the former by Caspar Bauhin in the passage quoted under the name of Alsines palustris minoris serpillifoliæ. It is depicted also by Johann Bauhin (but not described) in his Catalogue of the plants near Balneum Bollense [*Historia. . .fontis Bollensis,* 1598], p. 194.

(Facs. p. 94)

Explanation of the Names of Authors Cited

[*Note.* After each entry in Johnson's list is given, in square brackets, the full name, in an appropriate language, of the author concerned, together with the dates of his birth and death, where known, and the titles and dates of his work or works most likely to have been used by Johnson. I would like to acknowledge Professor F. A. Stafleu's and Mr Clive King's help with this section, J.S.L.G.]

Advers. Adversaria Petri Penae, & Matthiae de Lobell. [Pierre Pena (fl. 1530–1605) and Matthias de L'Obel (1538–1616); joint authors of *Stirpium Adversaria Nova*...London, 1570. Issued with *Altera Pars*, London, 1605.]

Anguil. Aloysius Anguillara. [Luigi Anguillara (d. 1570). *Semplici*... Vinegia, 1561.]

Bauh. Casparus Bauhinus. [Caspar Bauhin (1560–1624). *Pinax Theatri Botanici*, Basle, 1623. *Prodromus Theatri Botanici*, Frankfurt, 1620.]

Clus. Carolus Clusius. [Charles de L'Ecluse (1526–1609). *Rariorum Plantarum Historia*, Antwerp, 1601.]

Colum. Fabius Columna. [Fabio Colonna (? 1567–1650)...*Phytobasanos*..., Naples, 1592...*Ecphrasis*..., Rome, 1606.]

Cord. Valerius Cordus. [Valerius Cordus (1515–54). *Annotationes in P. Dioscoridis*...Strasburg, 1561.]

Diosc. Dioscorides. [Pedacius Dioscorides (fl. first century A.D.). *Opera*... Lyon and Frankfurt, 1598, and other editions.]

Dod. Rembertus Dodonaeus. [Rembert Dodoens (1517–85). *Stirpium Historiae Pemptades Sex*...Antwerp, 1583, and other works.]

Fuch. Leonardus Fuchsius. [Leonhard Fuchs (1501–66). *De Historia Stirpium*...Basel, 1542. *New Kreuterbuch*...Basel, 1543.]

Ger. Iohannes Gerardus. [John Gerard (1542–1612). *The Herball*..., London, 1597.]

(Facs. p. 95)

Ges. Conradus Gesnerus. [Konrad Gesner (1516–65). *Historia Plantarum*... Venice, 1541, and other works.]

Lob. Matthias de Lobell. [Matthias de L'Obel (1538–1616). *Plantarum*... *Historia*, Antwerp, 1576. *Icones*, Antwerp, 1581.]

Lugd. Historia Plantarum Lugduni cusa. [Published anonymously by Jacques Dalechamp (1513–88). The actual title begins *Historia Generalis Plantarum*...2 vols. Lyon, 1586–7. See Stafleu, 1967, p. 98.]

Matth. Petrus Andreas Matthiolus. [Pierandrea Mattioli (1501–77). *Opera*..., Basel, 1598.]

Offic. Officinis usitatum nomen. [Indicates a medicinal plant.]

Tab. Iacobus Theodorus Tabernaemontanus. [Jacob Theodor of Bergzabern (1520–1590), who wrote under the name 'Tabernaemontanus'. *Neuw Kreuterbuch*, 2 vols. Frankfurt, 1588–91.]

Thal. Iohannes Thalius. [Johannes Thalius (? 1542–83). *Sylva Hercynia*, published with *Hortus Medicus*...by J. Camerarius, Frankfurt, 1588.]

Trag. Hieronymus Tragus. [Hieronymus Bock (1498–1554), who wrote under the name 'Tragus'. *New Kreutterbuch*, Strasburg, 1539, and later editions.]

THE END

(Facs. p. 95)

Map of South-East England showing the County of Kent and the Borough of Hampstead.

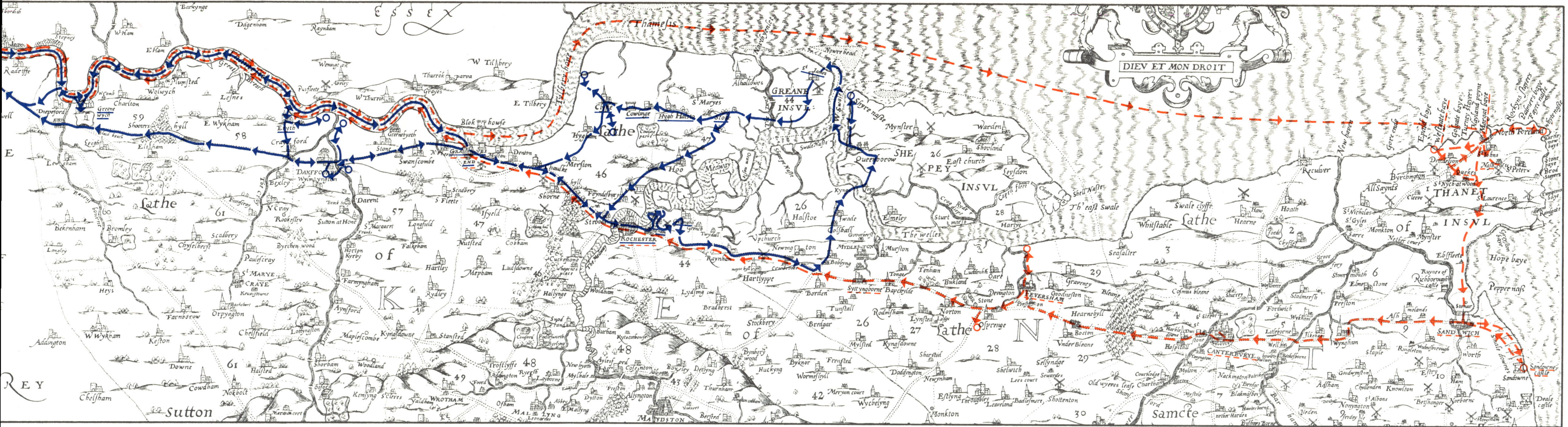
ESSEX
Thamesis
DIEV ET MON DROIT
GREANE
INSVL
SHEPEY
INSVL
THANET
INSVL
Lathe
Lathe of
Lathe
KENT
ROCHESTER
FAVERSHAM
CANTERBVRYE
SANDWICH
MAYDSTON
Sutton
Samcte

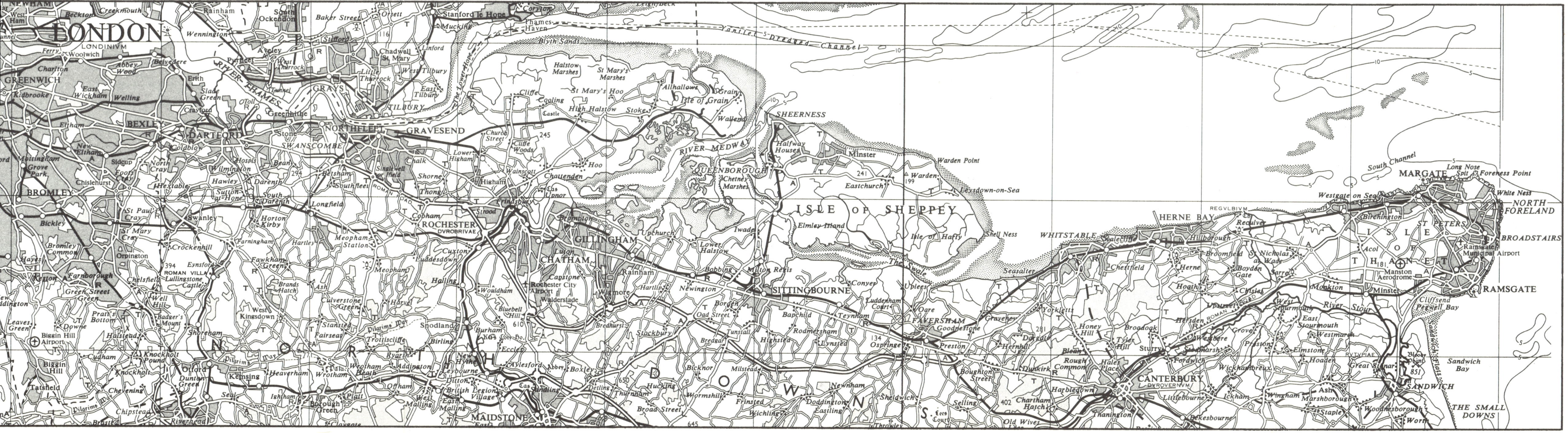
LONDON
LONDINIVM
GREENWICH
BEXLEY
DARTFORD
GRAVESEND
NORTHFLEET
SWANSCOMBE
TILBURY
GRAYS
RIVER THAMES
RIVER MEDWAY
SHEERNESS
QUEENBOROUGH
ISLE OF SHEPPEY
ROCHESTER
DVROBRIVAE
CHATHAM
GILLINGHAM
SITTINGBOURNE
FAVERSHAM
WHITSTABLE
HERNE BAY
MARGATE
NORTH FORELAND
BROADSTAIRS
RAMSGATE
ISLE OF THANET
CANTERBURY
DVROVERNVM
SANDWICH
MAIDSTONE
BROMLEY
NORTH DOWNS
THE SMALL DOWNS

Map of North Kent (Philip Symonson, 1596), showing the routes of Johnson's two journeys. The blue line shows the route described in the *Iter* (1629), and the red line that described in the *Descriptio* (1632). The small circles indicate the principal identifiable diversions from the main routes. Underlined towns are those mentioned in Johnson's accounts of the journeys. (Reproduced from the Ordnance Survey reprint of Symonson's map with the sanction of the Controller of H.M. Stationery Office, Crown Copyright reserved).

Modern map of North Kent (reproduced from the Ordnance Survey map with the sanction of the Controller of H.M. Stationery Office, Crown copyright reserved).

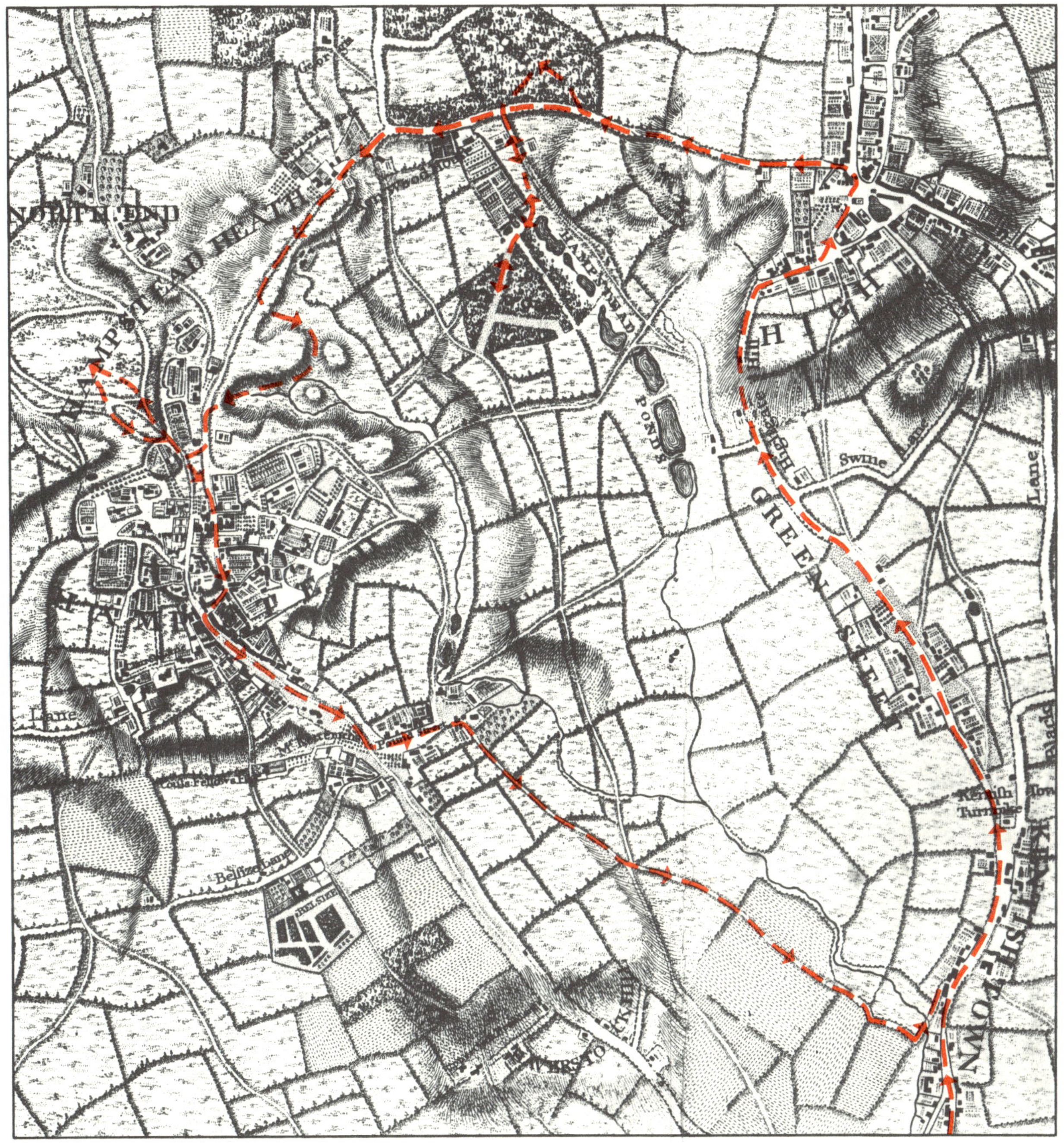

Map of Hampstead (John Rocque, 1746) showing Johnson's route described in the *Iter* (1629), (reproduced from the copy in the University Library, Cambridge, by permission of the Librarian).

Modern map of Hampstead (reproduced from the Ordnance Survey map with the sanction of the Controller of H.M. Stationery Office, Crown copyright reserved).

Appendices

I. Notes on Johnson's Names for Marine Algae

J. H. PRICE

II. Mr Duck's Sea Serpent

MAURICE BURTON

III. Notes on Johnson's Names for Bryophytes and Lichens

W. MARGADANT & P. W. RICHARDS

IV. Corrections and Additions to the original text of the *Iter* and the *Descriptio*

J. S. L. GILMOUR

APPENDIX I

Notes on Johnson's Names for Marine Algae

JAMES H. PRICE

1 *Bryon Lactucae foliis*, Lob. *Lactuca marina*, Taber. (see p. 104). *Bryon lactucae foliis* Plin. Lob. (see p. 59). *Lichen marinus* (see p. 59).

Johnson's (1633) revision of Gerard's *Herball* is a useful source of information regarding the concepts behind the names used both there and in the earlier Johnson publications. In the revised *Herball* Johnson both figures and comments at length on many of the 'species'. For *Lichen marinus* he gives an illustration which is taken directly from Lobel, 1576, *Plantarum seu stirpium historia*, where it appears on p. 647. This differs slightly from that given by Gerard in the 1597 version of the *Herball*, but both clearly relate to the same plant. The 1633 illustration appears on p. 1566, chapter 164, fig. 2, and is headed '*Lichen marinus*. Sea Lung-wort, or Oister-greene.' In 1632 Johnson himself gave *Lichen marinus*, Ger. as a synonym of *Bryon Lactucae foliis*, Lob. The 1629 *Iter* does not qualify *Lichen marinus* in any way with references or authors and unfortunately lists it immediately beneath and with the same status as *Bryon lactucae foliis* Plin. Lob. However it is very probable that *Lichen marinus* and *Bryon lactucae foliis* relate to one and the same plant, more especially since Ray (and Dillenius) in 1690 and 1724 equate *Oyster green* with *Lactuca marina*, with *Lichen marinus*, and with *Ulva marina Lactucae similis*.

I therefore conclude that both *Lichen marinus* (1629) and *Bryon lactucae foliis* Plin. Lob. (1629), together with *Bryon Lactucae foliis*, Lob., most probably all relate to the plant at present known as *Ulva lactuca* L.

2 *Alga membranacea ceranoides*, Bauh. (see pp. 104).

This name is not involved in the 1629 list but curiously appears twice in the 1632 list without significant difference. Johnson refers specifically to Bauhin in his use of this name and the latter (*Prodromus* 1620 and 1671, cap. VIII, entry VI, p. 155) gives a description which refers to the use of the Scots name Dils for his alga. On current applications of this local name only one of two genera could be involved—*Dilsea* or *Rhodymenia*, both being present in Kent nowadays. Luckily, Bauhin's morphological description is good enough to make it fairly certain that he is referring to *Rhodymenia*. Generally, this conclusion is supported by other subsequent authors who mention the alga. Ray, who in correspondence admitted his error, and Dillenius, 1724, after Ray had earlier (1670 and 1677) erroneously equated his own *Fucus sive Alga membranacea purpurea parva* (= ? *Porphyra*) with Bauhin's entity, separated the two and gave a description for the Bauhin name which agrees with *Rhodymenia*; Hudson (2nd and 3rd eds.) considers the Bauhin concept synonymous with Linnaeus' *Fucus palmatus*, which again leads back to *Rhodymenia*. Therefore, it seems tolerably certain that the species now known as *Rhodymenia palmata* (L.) Grev. is referred to.

3 *Fucus longissimo*, *latissimo*, *crassoque folio*..., and *Fuci ejusdem varietas polyschides* (see p. 104).

Although the name *polyschides* is now associated with the genus *Saccorhiza*, there is no indication from the figures given by Johnson of the characteristically knobbled and cushioned *Saccorhiza* holdfast, nor of the undulate stipe. Since *Saccorhiza* is now but dubiously known from Kent, it can be discounted as involved here. Indications from the figures would be that both stipes arising from the confluent holdfasts represent *Laminaria digitata*, especially since other parts of laminae appear behind the single one of fig. 1, without it being entirely clear whether the laminae are all part of the same plant or not. However, the legend which accompanies the figures states that, for fig. 1, '...unicum folium...' is present, whereas for fig. 2. '...Polyschides sive multifidum folium emitit, quod in 8. 9. 10 aut 12. plures, aut pauciores partes dividitur.' This distinction is confirmed and elaborated in English in the Johnson and Gerard *Herball* (1633). Since, on Thanet, confluent holdfasts with two or more emergent stipes of either *Laminaria digitata* only, or of mixed *L. digitata* and *L. saccharina*, are not at all uncommon, there seems little doubt left. I conclude that the legend and figure for fig. 1 relate to *Laminaria saccharina* (L.) Lamour., and those for fig. 2 to *Laminaria digitata* (Huds.) Lamour.

4 *Fucus marinus quartus*, Dod. (see p. 104).

This name is referred to directly by Johnson as a synonym of the fourth variety of *Quercus marina* which he recognizes in 1633. Luckily, he also gives an accurate illustration which leaves no doubt as to the attribution to be placed on the name. His concept is clearly that of the present *Halidrys siliquosa* (L.) Lyngb.

5 *Fucus spongiosus ramosus* (see p. 104).

Very little can be suggested for this name unless it be accepted without further proof that it is very likely near to *Fucus spongiosus nodosus* (see below). Dillenius (in Ray 1724, p. 49) comments correctly on the variation in size, degree of branching, colour, and shape in *Dumontia* and this has been our experience in Kent also. Thus a reasonable suggestion would be that this is merely another form of *Dumontia* which Johnson had mistaken for a different species. Since Johnson himself does not qualify his record, by text or illustration, and since no other authors have seen fit to comment on the record at all, nothing more can be said in the absence of specimens.

6 *Fucus spongiosus nodosus* (see p. 104).

This name, and the concepts behind it, present impossible complexities. The figure legend which appears in the 1632 *Descriptio* is repeated in English in the 1633 version of the *Herball* (p. 1569). The descriptive text in each of these cases fits reasonably well with the alga *Dumontia incrassata*, although the figure, which is also reproduced in 1633 (fig. 10, p. 1570), in Parkinson (1640, p. 1304, fig. 4), in Kew and Powell (1932, pl. IV), and in the 1847 reprint of the *Descriptio*, fig. 3, does not. Subsequent authors, including Dillenius in Ray, 1724, p. 49, no. 42 and Hudson, especially 1778 and 1798, p. 570, have clearly had this genus in mind (and have perhaps had additional specimens in front of them when writing the descriptions which they give) when they have cited *Fucus spongiosus nodosus* in their lists of synonyms. Dillenius gives a good description in which he indicates that there are profound faults in the figure given by Johnson in 1632 and 1633. This makes it a good deal easier to accept that probably all these authors are talking about *Dumontia* here; there remains the possibility that Johnson in fact had seen specimens of something else which is accurately pictured but badly described by him. If the latter is the case, he may have had either an old base of the brown alga *Cystoseira* (which is known as drift from the area), or of the brown alga *Halidrys*, or indeed some animal tissue as the basis of his record. Several authors have indicated subsequently that they believe this last possibility to be true. The point cannot be checked since specimens do not remain extant, so far as can be traced. *Dumontia* grows to a magnificent size in Thanet, but it is a plant which grows most luxuriantly in early season, February to April. To an extent, this would tally with the circumstances, since Johnson was there in August. It is almost inconceivable that, had he been there earlier, he could possibly have missed seeing the plant attached as it is then so widespread. In August, it is just possible that a specimen or two, in bad condition, could persist in the drift material, although I cannot remember ever seeing this genus as late in the year as that. An unusual set of seasonal sequences could easily account for that, however. In the circumstances, it is not unreasonable to attribute this record to *Dumontia incrassata* (O. F. Müll.) Lamour. However, the balance of evidence from all sources, including a large number of later authors' comments, seems to indicate that some colonial animal tissue may have been the primary source of the record.

7 *Fucus ferulaceus*, Lobelii (see p. 104).

Reference back to Lobel, 1576 (*Plantarum seu stirpium historia*, p. 653) reveals a figure which is clearly the original of that reproduced in Johnson and Gerard (1633, p. 1573, fig. 7,) although the latter has been turned on its side, perhaps as a space-saving measure. The figure in 1633 is associated with a description which indicates the similarity of the branching system with the leaves of the terrestrial *Foeniculum* (Fennel). Both the figure and this description tally well with the description and plates given by Roberts (*Br. phycol. Bull.* **3**, 547–64, 1968) for *Cystoseira foeniculacea* (L.) Grev. and do not differ markedly from facets of specimens held in BMNH. Unfortunately, we have never found specimens of any species of *Cystoseira* attached in Kent, although drift material of some species turns up occasionally, and field notes made by J. T. Neeve indicate that he found it attached at ELWS (extreme low water level of spring tides) on a rocky area just off Deal. I had thought at first that the species concerned was the red alga *Cystoclonium purpureum* (Huds.) Batt., but the swellings in that species are median in terminal branches, not beneath dichotomies as stated and figured by Johnson. Johnson and Gerard (1633, pp. 1572–3) record the habitat of the species they found as 'upon the rocks neere unto Narbone in France, and not far from the sea', which would fit the intertidal species *Cystoclonium* very well. They say nothing about the record being drift which, unless cataclysmic change has occurred, *Cystoseira* would certainly usually be in Kent. However, there is no doubt that much of the general evidence points in the direction of *Cystoseira foeniculacea*, which has been authentically recorded from as far east as Sussex and Hampshire and which is known to fruit in August in Brittany and elsewhere. It is just possible that Johnson's field identification and drawn figure are inaccurate to the degree that he in fact had *Cystoclonium*.

8 *Quercus marina vulgaris foliis non serratis* (see p. 104).

This fuller name seems to be the equivalent of the 1629 *Quercus marina* (see below). The full phrase is not used anywhere in Johnson and Gerard (1633) and no subsequent author seems to have referred to it. Johnson and Gerard (1633, pp. 1567–8) give several varietal names for *Quercus marina*, but the present name is not among them. Since the portion of the name after *marina* simply seems to be an emphasis of the common, non-serrate form of the frond, it is likely that Johnson was attempting more firmly to indicate the common form of his overall concept, before refining his ideas to the point of listing varieties as in 1633. Thus I conclude that this name also refers to *Fucus vesiculosus* L.

9 *Quercus marina foliis serratis* (see p. 104).

Not given elsewhere and neither illustrated nor given references to the authorities concerned. In view of the certain identity of *Quercus marina*, *sensu stricto*, however, and despite the fact that Johnson strangely did not include the present name as a variety in his 1633 revision, it is very unlikely that anything other than the modern *Fucus serratus* L. is referred to.

10 *Corallina ramosa parva*, and *pennata longior* (see p. 104).

With no amplifying data and few references by other authors, there is little that can be stated definitely. The most revealing information is given by Dillenius (in Ray, 1724, p. 33), who equates *Corallina Anglica* in Gerard's *Herball* (1597, p. 1379) with the *Corallina pennata longior* given by How (*Phyt. brit.*, p. 31, 1650). *C. Anglica* is figured on that page of Gerard and there is little doubt that it does represent a specimen of *Corallina officinalis* L. If Dillenius' synonymy was correct and the concepts of Johnson and How were equivalent, then the record for *C. pennata longior* probably relates to *C. officinalis* L. Dillenius (*loc. cit.*) also gives an entry for *C. ramosa parva*, in which he refers to a description in Plukenet's *Almagestum* and a figure in the *Phytographia* (fig. 3 of pl. 168) in the list of synonymy. This figure is poor, though very likely of a species of *Corallina*; it is not possible to say which species without material and I have not had time to check for the Plukenet specimens. For both these names, there is just the reservation that the areas characteristically associated with prolific growth of *Corallina* are not those cited by the above authors; even in the muddy habitats mentioned, however, there would presumably have been some artificial substrata present and it is by no means impossible that drift material was being considered. In summary, there is a fair possibility that *Corallina*, probably *officinalis*, is referred to here.

11 *Quercus marina* (see p. 59).

Gerard (1597, p. 1378, fig. 3) illustrates clearly *Fucus vesiculosus* under his *Quercus marinus* heading. This is maintained by Johnson and Gerard (1633, p. 1567, fig. 4) in that the later figure is clearly simply redrawn with a few modifications from the earlier. Clearly therefore, Johnson's concept of the name *Quercus marina(-us)* is the plant now known as *Fucus vesiculosus* L.

APPENDIX II

Mr Duck's Sea Serpent

MAURICE BURTON

We can either reject the story (p. 115) as a fake or accept it as a true record. We have no evidence for either other than this written account, and there is no reason to suppose that it was a fake. We can dismiss the idea of a sea serpent because, whatever may be the basis for that story, it has little to do with snakes (there are sea snakes in the Indian Ocean but they are not contributory to the sea serpent story). I would regard it as unlikely that the snake had escaped from a ship or a local zoo. Both of these could have been possible in the twentieth century, but there is nothing to indicate that they would have been likely to have occurred in the seventeenth century.

My considered suggestion is that this episode may be based on the finding of a very large grass snake (*Natrix natrix*). This snake is known to reach a length of $5\frac{1}{2}$ feet in this country and to exceed 6 feet in southern Europe. Such outsizes are rare, but there is no guarantee that they represent the maxima for this species. I have collected a few stories from countrymen telling of very large snakes in the English countryside and it is clear that when these are seen the witness does his best to put as much distance between the snake and himself as he can. My general conclusion would be therefore that there may well be rare instances of the grass snake attaining an even greater length than has been recorded in the scientific literature. Since even an average-sized grass snake will prey occasionally on small mammals I can well believe in the possibility of an outsize individual capturing rabbits, especially half-grown rabbits. Finally, the 15 feet may well be explainable not so much by a hypothetical outsize grass snake as the skin from, say, a 6 foot snake being stretched in the course of stuffing it with hay. That a snake's skin is highly extensible is shown by the bulge in a python that has swallowed an antelope. Experience in taxidermy suggests that a fresh skin, in unskilled hands, could undergo a hundred per cent increase, or more, in length and girth without any intent to deceive.

APPENDIX III

Notes on Johnson's names for Bryophytes and Lichens

WILLEM D. MARGADANT AND PAUL W. RICHARDS

The *Iter* and *Descriptio* record seven species of Muscus and two of Adiantum: all of these except Muscus clavatus (*Lycopodium clavatum* L.) refer to mosses or lichens. In Johnson's time the term 'muscus' was applied to lichens, hepaticae, and various other cryptogamic plants, as well as to mosses as now understood.

Since no specimens of mosses collected by Johnson exist and no descriptions are given with the names of the *Iter* and *Descriptio*, completely certain identification of his names is not possible and one can be guided only by his edition of Gerard's *Herball* (1633) and by other books available to him, and by the ecological probabilities.

Johnson's ideas on the nomenclature of mosses appear to be mainly derived from de l'Obel, from whose *Plantarum seu stirpium historia* the illustrations in his two printings (1633, 1636) of the *Herball* were copied. Useful clues to the identity of early moss and lichen names can be obtained from Dillenius' *Historia muscorum* (1741) which quotes large numbers of synonyms, including all the names used by Johnson. Dillenius' own names can be checked by the specimens in his herbarium, but of course his identification of earlier synonyms is not necessarily correct. In any case it has to be remembered that the lower plants had been studied only very superficially in Johnson's time and there is no evidence that he was particularly interested in them, so his names in many cases are likely to be group names rather than specific.

Hampstead Heath

Adiantum aureum maius

Dillenius (1741) gave Gerard, Petiver, Ray, etc. as authorities for this name and lists it as a synonym of Polytrichum quadrangulare vulgare Juccae foliis serratis. Hedwig (1801) gives the latter as a synonym of *Polytrichum commune*

Hedw., but *P. formosum* Hedw. was not distinguished from *P. commune* before Hedwig's time, so Johnson's name may well cover both species.

The figure of Muscus capillaris sive Adianthum aureum majus in Johnson's editions of Gerard might well represent almost any British species of *Polytrichum*, but the term 'maius' together with Dillenius' identification supports the view that one or both of the larger species *P. commune* and *P. formosum* was intended. It is interesting that Gerard remarks (referring apparently to A. aureum maius) (Johnson's Gerard, p. 1559), 'I found it in great abundance in a shadowie ditch upon the left hand neere unto a gate that leadeth from Hampstead Heath toward Highgate'. A ditch is a somewhat more likely habitat for *P. commune* than for any of the other species.

There is a later record for *P. commune* from Hampstead Heath and one from Ken Wood (Rose, 1964) for *P. formosum* which was also doubtless formerly common on the Heath.

Adiantum aureum minus

Though Dillenius identified this with his Bryum bulbiforme aureum [*Funaria hygrometrica* Hedw.], Gerard's figure hardly supports this and could represent almost any acrocarpous moss with a nodding capsule such as *Pohlia nutans* (Hedw.) Lindb., which was probably once common on the Heath, or some *Mnium* or *Bryum*. The remarks on page 1559 ('Of this Adiantum aureum there are three kinds, different only in magnitude and that the two bigger have many hairie threads upon their branches, where as the least hath only three or four close to the roote and is the least of plants that ever I saw') would also be compatible with the view that one of the smaller species of *Polytrichum* such as *P. juniperinum* or *P. piliferum* is intended.

Muscus pyxadatos alabastriculos imitatus, Lob.

Dillenius gave this name as a synonym of his varieties A and B of Coralloides scyphiforme, tuberculis fuscis. This must be *Cladonia pyxidata* (L.) Hoffm. or a closely related lichen species. Though probably now extinct, *C. pyxidata* was doubtless common on the Heath in earlier times.

Muscus terrestris vulgaris

The figure in Johnson's Gerard (taken from l'Obel) seems to represent a species of *Sphagnum* with capsules. Dillenius (1741) quoted this figure in the synonymy of his Sphagnum palustre molle deflexum, squamis cymbiformibus [*Sphagnum palustre* L.], though he gave Muscus terrestris vulgaris Ger. 'quoad descr.' as a synonym of a quite different moss, Hypnum dentatum vulgatissimum, operculis obtusis [*Brachythecium rutabulum* (Hedw.) Schimp.]. There are later records for other species of *Sphagnum* as well as for *S. palustre* from Hampstead Heath and it would be unwise to assume that Johnson distinguished between the various *Sphagnum* species.

Muscus in ericetis proveniens

L'Obel's figure, reproduced in Johnson's *Gerard*, is very difficult to interpret and looks like a poor copy by someone who did not understand the original. A description in Dutch by Dodonaeus suggests that it may represent a lichen such as *Cladonia coccifera* (L.) Willd., and Michelius also identified it as a lichen, but Dillenius believed it to be a moss and identified it as his Bryum stellare roseum pendulum minus, capsulis & setis longioribus [*Bryum capillare* Hedw.]. Certainly the upright stems, apparently branched above, could possibly be regarded as representing male shoots of an acrocarpous moss such as *Mnium hornum* L. or a *Bryum* sp. with conspicuous perigonia.

Muscus corallinus albicans, sive corallina, montana, Tab.

There can be little doubt that this was *Cladonia sylvatica* Hoffm. Dillenius gave 'Muscus corallinus albicans Johns. Plant. Hampst., p. 34', as a synonym of Coralloides montanum fruticuli specie, ubique candicans, 'The Ren Deer Moss, or common white Mountain Coralloides'. There are no recent records for Hampstead Heath but it is very likely to have grown there among *Calluna* which existed on the Heath in the earlier part of this century and was doubtless abundant there in Johnson's time.

Kent

Adiantum aureum majus, Tab. Polytrichum majus, Trag., Fuch.

As at Hampstead the moss referred to by Johnson was probably one of the two common large species of *Polytrichum*. Between Sandwich and Canterbury he is more likely to have encountered *P. formosum* Hedw. than *P. commune* Hedw., as the former might well have been found growing in a hedge or wood on clay-with-flints or some other decalcified material overlying the chalk. *P. commune* requires somewhat damper and more acid conditions than *P. formosum* and it is rather unlikely that suitable habitats for it existed in this part of Kent.

Adiantum aureum minus, polytrichum minus, Trag.

See above (p. 148).

Muscus capillaris altera species Dod.

Dillenius gave this in addition to Adiantum aureum minus as a synonym of Bryum bulbiforme aureum [*Funaria hygrometrica* Hedw.] and it is possible that the two names refer to different stages of development of the same species, e.g. before and after the elongation of the setae. It is not possible to tell whether Johnson was referring to *Funaria* or some other acrocarpous moss.

Muscus aquaticus denticulatus

Dillenius gave Muscus aquaticus denticulatus of Loesel (and other authors), as a synonym of Fontinalis triangularis major complicata, which is certainly *Fontinalis antipyretica* Hedw. It is curious however that he gave Johnson's Muscus major denticulatus, *Merc. Bot.* p. 52, as another synonym of the same species. Is one of these identifications wrong or did Johnson use two different names for the same moss? *Fontinalis antipyretica* is not rare in streams in Kent.

Muscus terrestris pyxidatus alabastriculos imitatus, Lob.

See above (p. 148).

APPENDIX IV

Corrections and Additions to the original text of the Iter *and the* Descriptio

J. S. L. GILMOUR

The Iter

In Goodyer's copy of the *Iter* in the Library of Magdalen College, Oxford, which Gunther (1922, p. 275) states 'appears to be a presentation copy to Goodyer' (see p. 7), the following MS. corrections are made, in Gunther's words, 'by the author(?)':

p. 33, [1] l.30: for festinantes read festinantibus
p. 33, l.32: for *Glycyrhiza* read *Glycyrrhiza*
p. 34, l.23: for audeo read ausim
p. 34, l.23: for possum read possim
p. 36, l.5: for est read sit
p. 36, l.11: for investigemur read investigemus
p. 37, l.29: for possumus read possimus
p. 39, l.4: for potuimus read potuerimus
p. 41, l.31: after *Androsaemum quorundam* add Park-leaves
p. 43, l.3: after sit add quod

On $[B_4]^r$ (p. 43) of the copy of the *Iter* at Magdalen College there is a MS. list of the names of 38 plants. These are printed by Gunther (1922, p. 275), who suggests that they were added by Goodyer. No indication is given of the purpose of the list but, from the species included, they were probably intended as additions to those mentioned in the *Iter*, though there is a small overlap.

The Descriptio

In the separately bound copy of the *Descriptio* at Magdalen College, Oxford (see p. 7), which Gunther (1922, pp. 232 and 275) suggests was successively owned by Johnson, William How, and Goodyer, there are the following MS.

[1] The page numbers are those of the continuous pagination of this book.

alterations and additions, some of them of very great interest (see Gunther, 1922, p. 232):

(1) A number of alterations and additions to the text, which Gunther states are in Johnson's handwriting. Gunther prints these in full, and notes that the additions are all included in How's *Phytologia* (1650). The most interesting species added (on the blank page opposite the list of Nominum... Authorum... p. 95, right, of continuous pagination) is perhaps 'Trifolium pumilum supinum flosculis albis...Angl. White Dwarf Trefoile' [*Trifolium subterraneum* L.], the earliest printed British record for which (Clark, 1900, p. 37) is in Johnson's *Mercurius*, 1634, p. 73.

(2) Eighteen leaves are bound in at the end of the volume, bearing the following MS. material.

Ff. 1 and 2. Blank.

Ff. 3–6. An index to the genera included in the book, stated by Gunther to be in Johnson's hand, with additions by How.

F. 6 verso. Notes by How.

Ff. 7–11. Alphabetical list, in How's hand, of about 190 British species not included in Johnson's lists.

Ff. 12–17. recto. Blank.

Ff. 17 verso, and 18. Notes by How.

For a fuller account and an illustration of these MS. additions see Gunther (1922), pp. 232 and 276–8. As Gunther points out, 'this little volume has the great sentimental interest of being the germ from which all British Floras are descended', since it was used by How as a basis for his attempt at a complete account of British plants – the *Phytologia* of 1650.

References

The following list includes all publications cited in the text, but omits a number of standard reference works and library catalogues which have been consulted, and also most of the authorities cited for Johnson's life by Kew and Powell (1932, pp. 3–10) in their bibliography. The works included in Johnson's 'Explanation of the names of authors cited' (see the translation of the *Descriptio*, p. 134) have not been repeated in this list, unless quoted elsewhere.

ARBER, AGNES R. (1912) – *Herbals, their origin and evolution. A chapter in the history of Botany 1470–1670.* Cambridge, University Press.

ARBER, EDWARD (1875–94) – *A list, based on the registers of the Stationers' Company, of 837 London publishers between 1553 and 1640 (transcript).* 5 vols. Birmingham [Privately printed].

BARRATT, THOMAS J. (1912) – *The annals of Hampstead.* 3 vols. London, A. & C. Black.

BARRETT, C. R. B. (1905) – *The history of the Society of Apothecaries of London.* London, Elliot Stock.

BAUHIN, CASPER (1620) – *Prodromus theatri botanici.* Frankfurt, Typis P. Jacobi, impensis I. Treudelii. Another edition 1671.

BECKER, M. JANET (1930) – *Rochester bridge: 1387–1856. A history of its early years, compiled from the wardens' accounts.* London, Constable & Co. Ltd.

BOULGER, GEORGE SIMONDS (1892) – 'Thomas Johnson' in *Dictionary of national biography.* London.

BRUCE, J. AND HAMILTON, W. D. (eds.) (1859) – *Calendar of state papers domestic 1628–9.* 23 vols. London, Longman & Co. (1858–97).

Calendar of state papers domestic (1858) – See Lemon, R. (ed.).

Calendar of state papers domestic (1859) – See Bruce, J. and Hamilton, W. D. (eds.).

CAMDEN, WILLIAM (1722) – *Britannia; or, a chorographical description of Great Britain and Ireland, together with the adjacent islands.* Ed. 2, revised by Edmund Gibson. London, Printed by M. Matthews for A. Churchill.

CLAPHAM, A. ROY, TUTIN, THOMAS G. AND WARBURG, EDMUND F. (1962) – *Flora of the British Isles.* Ed. 2. Cambridge, University Press.

CLARKE, WILLIAM AMBROSE (1900) – *First records of British flowering plants.* Ed. 2. London, West, Newman & Co.

CORBETT, SIR JULIAN STAFFORD (1904) – *England in the Mediterranean; a study of the rise and influence of British power within the Straits, 1603–1713.* London, New York, and Bombay, Longmans, Green and Co.

DE CRESPIGNY, EYRE CHAMPION (1877) – *A new London Flora; or Handbook to the botanical localities of the Metropolitan Districts.* London, Hardwicke & Bogue.

DILLENIUS, JOANNES JACOBUS (1724) – *See* Ray, John (1724).

(1741) – *Historia muscorum.* Oxford, Theatro Sheldoniano.

GERARD, JOHN (1597) – *The Herball, or generall historie of plantes.* London, J. Norton.

GERARD, JOHN (1633) – *The Herball or generall historie of plantes*. Ed. 2 enlarged and amended by T. Johnson. London, A. Islip, J. Norton and R. Whitakers. Reprinted in 1636.

GUNTHER, ROBERT THEODORE (1922) – *Early British botanists & their gardens*. Oxford, Printed by F. Hall for the author at the University Press.

(1925) – *Early science in Oxford*. 5 vols. Oxford [Printed for subscribers].

Hampstead Heath, its geology and natural history (1913). London & Leipzig, Hampstead Scientific Society.

HANBURY, FREDERICK JANSON AND MARSHALL, EDWARD SHEARBURN (1899) – *Flora of Kent with notes on the topography, geology, and meteorology of the county*. London, F. J. Hanbury. Abbreviated Fl.K. in text.

HASTED, EDWARD (1797–1801) – *The history and topographical survey of the county of Kent, containing the antient and present state of...it...collected from public records, and illustrated with maps...views, etc.* Ed. 2. 12 vols. Canterbury, Simmons and Kirkby.

HEDWIG, JOHANN (1801) – *Species muscorum frondosorum descriptae et illustratae*. Leipzig, Joannis Ambrosii Barth.

HOW, WILLIAM (1650) – *Phytologia Britannica, natales exhibens indigenarum stirpium sponte emergentium*. London, Typis Ric. Cotes, impensis Octaviani Pulleyn.

HUDSON, WILLIAM (1762) – *Flora Anglica, exhibens plantas per regnum Angliae sponte crescentes*. London, J. Nouraw, C. Moran.

JEFFERS, ROBERT H. (1967) – *The friends of John Gerard*. Falls Village, Conn., Herb Grower Press.

JOHNSON, THOMAS (1629) – *Iter plantarum investigationis ergo susceptum a decem socii in agrum Cantianum 1629...ericetum Hamstedianum sive Plantarum ibi crescentium observatio habita anno eodem*. [Printer unknown].

(1632) – *Descriptio itineris plantarum investigationis ergo suscepti in agrum Cantianum 1632 et enumeratio plantarum in ericeto Hampstediano locisqe vicinis crescentium*. London, Tho. Cotes.

(1633) – *See* Gerard, John (1633).

[transl.] (1634) – *The workes of...Ambrose Parey. Translated out of Latine by T. J. and compared with French*. London, Th. Cotes and R. Young.

(1634) – *Mercurus botanicus; sive plantarum gratia suscepti itineris, anno 1634 descriptio*. London, T. Cotes.

(1634) – *Thermae Bathonicae*. London (published with the above).

(1639) – *Joannis du Boys observationes in methodum miscendorum medicamentorum topicorum*. Edited by T. J. Londini, Typis Edwardi Griffini sumptibus Richardi Whitakeri.

(1641) – *Mercurii botanici pars altera, sive plantarum gratia suscepti itineris in Cambriam ...descriptio. Exhibens reliquarum stirpium nostratium...catalogum*. London, Excudebat T. & R. Cotes.

KENT, DOUGLAS HENRY AND LOUSLEY, J. EDWARD (1951–7) – *A hand list of the plants of the London area: flowering plants, ferns and stoneworts*. London.

KEW, HARRY WALLIS AND POWELL, H. E. (1932) – *Thomas Johnson, botanist and royalist*. London, Longmans, Green & Co.

LEMON, R. (ed.) (1858) – *Calendar of state papers domestic 1619–23.* Addenda by M. A. E. Green. 7 vols. London, 1856–71.

LOBEL, MATHIAS DE (1576) – *Plantarum seu stirpium historia.* Anvers, C. Plantin.

McKERROW, RONALD B. (1913) *Printers' and publishers' devices in England and Scotland 1485–1640.* London. Printed for the Bibliographical Society at the Chiswick Press.

OPPENHEIM, MICHAEL (1896) – *A history of the administration of the royal navy and of merchant shipping in relation to the navy, from MDIX to MDCLX with an introduction treating of the preceding period.* London & New York, J. Lane.

PARKINSON, JOHN (1629) – *Paradisi in sole paradisus terrestris: or a Garden of all sorts of pleasant flowers which our English ayre will permitt to be noursed up: with a kitchen garden and an orchard together with the right orderinge, planting & preserving of them.* London, H. Lownes and R. Young.

(1640) – *Theatrum botanicum: the theater of plants; or, an herball of a large extent.* London, T. Cotes.

Phytologist, The (1848). Review of Ralph's *Opuscula,* signed L., pp. 114–124. London.

PLUKENET, LEONARD (1691–2) – *Phytographia, sive stirpium illustriorum & minus cognitarum icones.* London, Sumptibus autoris.

(1696) – *Almagestum botanicum, sive phytographiae plucnetianae onomasticon methodo synthetica digestum...adjiciuntur & aliquot novarum plantarum icones.* London, Sumptibus autoris.

RALPH, THOMAS SHEARMAN (1847) – *Opuscula omnia botanica T. Johnsoni nuperrime edita a T. S. Ralph.* London, Guliel. Pamplin.

(1849) – *T. Johnsoni descriptio itineris plantarum...Nup. edit. T. S. Ralph.* London, Guliel. Pamplin.

RAVEN, CHARLES EARLE (1940) – *John Ray, naturalist, his life and works.* Cambridge, University Press.

(1947) – *English naturalists from Neckam to Ray; a study of the making of the modern world.* Cambridge, University Press.

RAY, JOHN (1670) – *Catalogus plantarum angliae, et insularum una cum observationibus & experimentis novis medicis & physicis.* London, Typis A. Clark, impensis J. Martyn. Ed. 2, 1677.

(1690) – *Synopsis methodica stirpium Britannicarum, in qua tum notae generum characteristicae traduntur, tum species singulae breviter describuntur.* London, S. Smith.

(1696) – *Editio secunda* [of Synopsis] *acessit a Rivini epistola ad J. Rouim de methodo: cum ejusdem responsoria, in qua D. Tournefort elementa botanica tanguntur.* London, S. Smith and B. Walford.

(1724) – *Editio tertia* [of Synopsis] (Edited by J. J. Dillenius). London, G. and J. Innys.

SMITH, FREDERICK FRANCIS (1928) – *A history of Rochester.* London, C. W. Daniel Company.

STAFLEU, FRANS A. (1967) – *Taxonomic literature: a selective guide to botanical publications with dates, commentaries and types.* Utrecht, International Bureau for Plant Taxonomy and Nomenclature.

STOW, JOHN (1615) – *Annales, or, a generall chronicle of England.* (Begun by John Stow and finished by Edmund Howes.) London, T. Dawson for T. Adams.

TAIT, G. A. (1909) – *The church & vicars of Dartford.* Dartford.

TRIMEN, HENRY AND DYER, W. T. THISTLETON (1869) – *Flora of Middlesex: a topographical and historical account of the plants found in the county; with sketches of its physical geography and climate and of the progress of Middlesex botany during the last three centuries.* London, R. Hardwicke. Abbreviated Fl.M. in text.

UNDERWOOD, E. ASHWORTH (1962) – *A history of the Worshipful Society of Apothecaries* (Vol. 1 of Wellcome Historical Med. Mus. Publications). London.

VENN, JOHN AND J. A. (1922) – *Alumni Cantabrigienses; a biographical list of all known students, graduates and holders of office at the University of Cambridge, from earliest times to 1900.* Cambridge, University Press.

WALKER, JOHN (1714) – *An attempt towards recovering an account of the numbers and suffering of the clergy of the church of England who were sequester'd harassed etc... in the ... times of the great rebellion, occasion'd by the ninth chapter... of Dr. Calamys Abridgement of the life of Mr. Baxter.* London, unknown or J. Nicholson, etc.

Indexes

1. General
2. Index of Names of Plants & Animals

1. General Index

This index covers the persons and places mentioned in the *Iter* and the *Descriptio*, together with certain items from the Introduction and Appendices which it was felt might be of use to readers.

2. Index of Names of Plants & Animals

This index includes the modern scientific names of the plants and animals mentioned in the *Iter* and the *Descriptio*, and in the Introduction and Appendices, together with a few English names used by Johnson. Johnson's Latin names are not included, but these can, of course, be found by looking up the modern name in the appropriate pages of the translation.